T0133027

Contesting Medical Confidentiality

Contesting Medical Confidentiality

Origins of the Debate in the United States, Britain, and Germany

Andreas-Holger Maehle

The University of Chicago Press
Chicago and London

The University of Chicago Press, Chicago 60637
The University of Chicago Press, Ltd., London

Printed in the United States of America

25 24 23 22 21 20 19 18 17 16   1 2 3 4 5

ISBN-13: 978-0-226-40482-0 (cloth)
ISBN-13: 978-0-226-40496-7 (e-book)
DOI: 10.7208/chicago/9780226404967.001.0001

Library of Congress Cataloging-in-Publication Data
Names: Maehle, Andreas-Holger, 1957– author.
Title: Contesting medical confidentiality : origins of the debate in the United States, Britain, and Germany / Andreas-Holger Maehle.
Description: The University of Chicago Press : Chicago ; London, 2016 | Includes bibliographical references and index.
Identifiers: LCCN 2016008497| ISBN 9780226404820 (cloth : alk. paper) | ISBN 9780226404967 (e-book)
Subjects: LCSH Confidential communications—Physicians—History—19th century. | Confidential communications—Physicians—History—20th century. | Confidential communications—Physicians—United States—History. | Confidential communications—Physicians—Great Britain—History. | Confidential communications—Physicians—Germany—History. | Physicians—Professional ethics. | Privacy, Right of—History.
Classification: LCC K3611.C65 M34 2016 | DDC 344.03/21—dc23 LC record available at http://lccn.loc.gov/2016008497

♾ This paper meets the requirements of ANSI/NISO Z39.48-1992 (Permanence of Paper).

Contents

Introduction

Medical confidentiality is widely seen as the cornerstone of an effective physician–patient relationship. Patients' assumption that the personal information they give to doctors will be kept in confidence encourages them to be open about the details and circumstances of their illnesses. This personal information aids doctors in making correct diagnoses and prescribing appropriate treatments. Keeping sensitive knowledge about patients secret is furthermore regarded as an expression of respect for their privacy.[1] This is the ideal picture. Often the precept of secrecy in the Hippocratic Oath is cited in this context in order to emphasize the ancient authority of this rule for the medical profession.[2] History, however, teaches a different lesson. Medical secrecy was, and remains, a controversial subject. Doctors have breached confidentiality in the interest of persons other than the patient, or they have been forced by the state to disclose information about patients in the name of the law or of public health. Simultaneously, secrecy has assiduously been guarded by the medical profession. The following examples may briefly illustrate different types of problems with medical confidentiality.

In June 1870, the Committee on Ethics of the New York Academy of Medicine found one of their fellows, the world-renowned surgeon Dr. J. Marion Sims (1813–83), guilty of having betrayed the secrets of a patient as well as having resorted to public advertisement, and it recommended that the penalty of a "Reprimand of the Offender by the President of the Academy" be administered without delay.[3] Sims's offence consisted in having written a (subsequently published) letter to the editor of the *New York Times*, in which he detailed the conservative advice he had given to the famous actress Charlotte Cushman (1816–76) when she consulted him with worries about a "little indurated gland." Contrary to his advice, Miss Cushman had ("unfortunately," as Sims noted) the gland extirpated while she was in Britain. The operation was followed by some complications: in an article published a couple of weeks before Sims's letter, the newspaper's foreign correspondent had reported that Cushman was lying "at the point of death, quite given up by her friends." Sims expressed his opinion that the complication must have been "erysipelas" (that is, an inflammation of the skin and underlying soft tissues), which would be self-limiting in a short time, terminating either in convalescence or in death. Since no news of Miss Cushman's death had reached New York, Sims encouraged the readers to assume that she was recovering.[4]

Sims's assessment proved right in that the actress did recover from the operation (though she died some years later of cancer, of which the "indurated gland" had probably been an early sign).[5] Yet a Dr. Richardson subsequently filed charges against Sims with the secretary of the academy's ethics committee, Dr. Thomas C. Finnell (1826–90), accusing the surgeon of breach of medical confidentiality and of advertising his expertise through association with his prominent patient. Such conduct, the charges claimed, violated two paragraphs in the Code of Ethics of the American Medical Association, which the New York Academy had adopted—namely, the requirement of secrecy and the prohibition of public advertising.[6] Sims was given the opportunity to provide a written defense, which he did, denying any attempt to advertise his skills and maintaining that he had written the letter to relieve the anxieties of the actress's many friends.[7] However, the Committee on Ethics was not convinced and, as mentioned, found Sims guilty on both counts.

Sims's case was more complex than this brief summary can convey, not least through the double charge of breach of confidentiality and public advertising. The latter also referred to journalists' newspaper articles on his work—in particular, his novel technique for operating on vesicovaginal fistulas, which he had demonstrated to eminent surgeons in Paris and for which the French government had awarded him the decoration of the *Légion d'Honneur*. Also, the surgeon who had operated on Miss Cushman was the equally famous Sir James Young Simpson (1811–70) in Edinburgh.[8] But issues of professional jealousy or rivalry aside, the seriousness with which the New York Academy of Medicine dealt with the charges illustrates how important confidentiality was for the reputation of medical men. Even for a fellow as prominent and admired as Sims, no exceptions were made: Sims's breach of confidentiality—whether made in the interest of others, as he claimed, or from negligence, or disregard of the professional code of ethics—could not go unpunished. The discipline of the organized medical profession demanded this.[9]

Another example, this time from imperial Germany, sheds light on the tension between the state's claim on medical information for the prosecution of criminal offences and doctors' obligation of confidentiality. In 1910, in the wake of workers' street riots in Berlin-Moabit, the police confiscated medical case notes after the doctors concerned had refused to give information about any injured persons they might have treated. Ludwig Ebermayer (1858–1933), a lawyer at the German Supreme Court (*Reichsgericht*), defended the police's action in the medical professional press. He argued that the doctors should have voluntarily disclosed the relevant information, since the security of the state and the public good had been in danger, and they should therefore have followed a higher moral duty that superseded medical confidentiality.[10] Albert Hellwig (1880–1950), a Berlin lawyer, took a similar line, pointing out that medical practitioners should respect such higher interests, because they were "in the first instance" citizens with duties to the community, not just to their individual patients.[11] Breaches of professional secrecy were punishable under the Reich Penal Code of 1871, and since the late 1870s, German doctors could claim medical privilege in court proceedings—that is, they had a legal entitlement to refuse to testify (*Zeugnisverweigerungsrecht*), similar to Catholic priests' recognized right to withhold information obtained during confession. Doctors were wrong, however, in assuming that they were entitled to treat patient information not only as confidential but also as "privileged" in the sense that they could generally withhold it from the authorities. Patients in Germany could not expect such *absolute* secrecy from their doctors. Only in 1953, through a change in criminal law in Germany (*Strafrechtsänderungsgesetz*), was a regulation introduced that prohibited the confiscation of medical case records unless the patient had released the doctor from his duty of confidentiality.[12] However, regardless of this prohibition (under section 97 of the Code of Criminal Procedure), in 1986, tax authorities seized a total of 1,348 patient file cards from the practice of a gynecologist in Memmingen, Bavaria, who was suspected of tax evasion by failing to declare income from cash payments received for carrying out abortions. Numerous women identified through these case notes were prosecuted and punished for illegal abortion and questioned as witnesses during the criminal proceedings against the doctor. The case led to considerable public debate in West Germany. The Federal Supreme Court (*Bundesgerichtshof*) held in 1991 that the confiscation of patient records in a criminal investigation of a doctor for illegal abortions was lawful. In 2000, a constitutional complaint by the Memmingen doctor was not accepted for decision by the Federal Constitutional Court (*Bundesverfassungsgericht*).[13] The more far-reaching question, however, of whether doctors should set aside medical confidentiality to assist the authorities in the prosecution or prevention of crimes, continues to be an issue of legal debate, not only in Germany, but also in the United States, Britain, and France.[14]

These examples of, on the one hand, a doctor's breach of his professional duty of confidentiality and, on the other hand, doctors being forced to reveal patient information to the authorities, illustrating the limits of their privilege, reflect different key aspects of the major theme of this book: how doctors, lawyers, and the wider public have understood and negotiated the boundaries of medical secrecy. In particular, I will examine the conflict between individual interests in confidentiality and collective interests in disclosure by analyzing three historical debates: about the privilege of doctors to be exempt from the duty of testimony in court; about the responsibility of doctors to breach patients' confidence in the interest of fighting the spread of venereal diseases; and about a possible duty of doctors to report cases of illegal abortion to the authorities. There were lively debates on all three issues in the United States as well as in Britain and Germany during the late nineteenth and early twentieth centuries that have allowed me to present my historical findings in a three-country comparison. While my overall narrative reaches from late eighteenth-century England to post–World War II Germany, my comparison focuses on the period from the 1890s to the 1920s, when positions of the medical profession, the legal profession, and the state clashed over those questions of confidentiality. There were, of course, further areas of concern about medical secrecy, for example, in relation to life insurance, psychiatric patients, and clinical teaching.[15] However, these areas were addressed only sporadically and did not produce the amount of comment and controversy that the issues of medical privilege, venereal disease, and abortion caused.

Several authors have taken a historical perspective on medical confidentiality before me, but they have largely examined the topic in only one national context. For instance, Raymond Villey has explored the French tradition of medical secrecy, which for much of the nineteenth century was conceived as "absolute."[16] Amy Fairchild, Ronald Bayer, and James Colgrove have discussed programs of disease surveillance in twentieth-century America with regard to their implications for patient privacy, including the controversies surrounding the control of AIDS.[17] Recently, Angus Ferguson has provided the first comprehensive study of the development of medical confidentiality in Britain from the eighteenth to the late twentieth century.[18] There are also several substantial articles and book chapters on specific aspects of, or cases relating to, the history of medical confidentiality.[19] Readers interested in the details of health policies, programs, guidelines, and administrative processes, and in the personalities involved, will find here rich sources of information. In my comparative study, I invite the reader on a journey through three historical debates on medical confidentiality in three countries in order to discover how different (or similar) solutions have been sought for a continuing problem: the conflict between individual and collective interests in the question of secrecy versus disclosure. In this way, I wish to make a contribution to the history of medical ethics and law that will speak to today's ongoing concerns about patient privacy, for example, regarding the protection of health data.[20] As the reader will find, no clear-cut or simple lessons can be drawn from the historical record. Confidentiality was (and is) an issue deeply entangled with other concerns, such as the reputations of doctors as well as of patients, the health of the population, and the promotion of justice in legal proceedings. Not least, the contexts of health care have changed significantly since the early twentieth century so that no easy inferences from the past to the present can be made. Still, I hope my historical account and analysis may help readers understand the basic ethical issues at stake in medical confidentiality that continue to give rise to public debate.

Chapter 1

Medical Privilege in Court: Protecting Patient Confidence or Obstructing the Course to Justice?

Introduction

One of the main problems for medical confidentiality in the nineteenth and early twentieth centuries was the question of whether doctors could be required to give evidence in court about their patients' physical or mental conditions. The ethical duty of medical secrecy, already expressed in the Hippocratic Oath around 400 BC,[1] was widely regarded as constitutive of the physician–patient relationship, as it would support the trust patients placed in their doctors. This trust would encourage a patient to reveal details of his or her condition and its history that would help the doctor in arriving at the right diagnosis and in choosing the appropriate treatment. Such benefits of confidentiality were seen in some jurisdictions as a reason to exempt doctors from testifying to private details of their patients' medical information in court. Knowledge that personal medical information might be divulged in open court might prevent patients from seeing doctors in sensitive cases of illness, to the detriment of their health as well as of public health. On the other hand, valuable evidence might be lost through exclusions of medical testimony, perhaps even leading to judicial errors. This conflict—between the court's mission to establish the truth and the desire to protect patients' beneficial, fiduciary relationship to their doctor—was at the heart of many debates on medical confidentiality. It was controversial whether a medical privilege in court—that is, a physician's or surgeon's right to refuse giving evidence—was justifiable in the same way as the recognized legal privilege that protected the communications between attorney and client. Moreover, it was questionable whether the relationship between doctor and patient was truly comparable with that between a priest and his penitent and whether medical secrecy should be treated with the same respect as the seal of confession. This chapter discusses the different approaches to tackling the problem of medical confidentiality in court in Britain, the United States, and Germany and highlights the various arguments adduced by legal and medical commentators.[2]

Legal Preconditions in Britain

The question of a medical privilege first arose publicly in late eighteenth-century England in the trial for bigamy of Elizabeth Chudleigh (1720–88), Duchess of Kingston.[3] During this trial, held in April 1776 in front of the House of Peers, the duchess's surgeon and friend, Caesar Hawkins (1711–86), was asked by counsel for the prosecution whether he had known of a previous marriage between her and the naval officer August John Hervey, who had since become Earl of Bristol. Hawkins, who had been present at the birth of Chudleigh's and Hervey's child and had attended to the child before it died in infancy, was reluctant to answer the question. Instead, he raised the issue of medical confidentiality by repeatedly saying, "I do not know how far any Thing, that has come before me in a confidential Trust in my Profession, should be disclosed, consistent with my professional Honour."[4] In response, Lord Chief Justice Mansfield (William Murray, 1705–93) made a statement that would set a precedent for centuries to come:

> If no Lord differs in opinion, but thinks that a Surgeon has no Privilege to avoid giving Evidence in a Court of Justice, but is bound by the Law of the Land to do it [. . .] if all your Lordships acquiesce, Mr. *Hawkins* will understand, that it is your Judgment and Opinion, that a Surgeon has no Privilege, where it is a material Question, in a

> Civil or Criminal Cause, to know whether Parties were married, or whether a Child was born, to say, that his Introduction to the Parties was in the Course of his Profession, and in that Way he came to the Knowledge of it. [. . .] If a Surgeon was voluntarily to reveal these Secrets, to be sure he would be guilty of a Breach of Honour, and of great Indiscretion; but, to give that Information in a Court of Justice, which by the Law of the Land he is bound to do, will never be imputed to him as any Indiscretion whatever.[5]

None of the Lords objected, and Hawkins subsequently gave evidence. The Duchess of Kingston was eventually found guilty of bigamy and "demoted" to Countess of Bristol. However, as a peeress, she was spared the usual corporal punishment for bigamy of being branded on the thumb, and before legal action to contest the will of the late Duke of Kingston could start, she had escaped to France.[6]

With Lord Mansfield's statement, a privilege for medical men to refuse giving evidence about their patients, analogous to lawyers' established right to silence about their conversations with clients, had been rejected in the highest English court. Occasionally, judges lamented the fact that the law of privilege did not include medical practitioners. Justice Buller, for example, remarked during a trial in 1792 (*Wilson v. Rastall*), "There are cases, to which it is much to be lamented that the law of privilege is not extended: those in which medical persons are obliged to disclose the information, which they acquire by attending in their professional characters. This point was very much considered in the duchess of Kingston's Case, where sir C. Hawkins, who had attended the duchess as a medical person, made the objection himself, but was over-ruled, and compelled to give evidence against the prisoner."[7] However, Buller said this as an aside when the actual issue at stake in the case, which was about bribery of voters in a borough election, was the confidentiality of communications between attorneys and their clients, not between medical men and their patients. Lord Mansfield's opinion in the Duchess of Kingston case was adopted in most English courts, turning it into a principle of common law.[8] It also became accepted in Scots law, particularly after Lord Fullerton, as one of the judges in a case heard in the Scottish Court of Session (*AB v. CD* 1851), had endorsed it.[9] In this case, a doctor was sued by a kirk elder for a breach of medical confidentiality by disclosing sensitive family information to the minister of the parish. The information suggested that the kirk elder's child had been conceived before marriage, and the elder had therefore been dismissed from the kirk session. Lord Fullerton confirmed the defense lawyer's point, referring to the trial of the duchess, that there was no medical privilege in court. On the other hand, this Scottish case established the principle that secrecy was an integral part of the contract between a medical man and his client.[10]

It has been argued that Caesar Hawkins's reluctance in the Duchess of Kingston case to testify to the personal circumstances of his prominent patient predominantly had to do with his desire to be recognized as a man of honor—a gentleman—which was crucial for his professional reputation and the success of his practice.[11] In fact, contemporary writers on medical ethics, such as the Edinburgh professor of medicine John Gregory (1724–73) and the Manchester physician Thomas Percival (1740–1804), included their demands for discretion in a framework that sought to establish the gentlemanly and professional conduct of physicians and surgeons. A physician, as Gregory had pointed out in 1772, often learned through his profession of the "private characters and concerns of the families" in which he was employed and got to see people "in the most disadvantageous circumstances" and "humiliating situations." His patients' reputations might thus depend on his "discretion, secrecy, and honour." "Secrecy," Gregory emphasized, "is particularly requisite where women are concerned."[12]

Similarly, Percival exhorted the readers of his book *Medical Ethics* (1803) that "the familiar and confidential intercourse, to which the faculty [that is, medical profession] are admitted in their professional visits, should be used with discretion and with the most scrupulous regard to fidelity and honour." In his view, "secrecy and delicacy, when required by peculiar circumstances, should be strictly observed."[13] For Percival, however, confidentiality does not appear to have been an *absolute* duty. Rather, medical secrecy, in his understanding, spared patients embarrassment in particular situations. Adherence to secrecy was, for him, one of the professional duties required by medical ethics. Observance of these duties formed the character of physicians and surgeons as gentlemen.[14] In his discussion of their role as expert witnesses in court, Percival reminded physicians of their legal obligation to say "*the truth, the whole truth,* and *nothing but the truth,*" even if their testimony might lead to capital punishment of the defendant.[15] He did not address the question of medical privilege in this context. Twenty years later, in the case of *Rex v. Elizabeth Gibbons* (1823), a surgeon, Mr. Cozzens, who had attended to the defendant while she was on trial for having killed her illegitimate child, was held by the judge not to be entitled to refuse testimony concerning her confession to him. The Duchess of Kingston case served as a precedent for this ruling.[16]

Medical Privilege in the United States

In Britain, the rejection of a medical privilege in court seems—initially—to have met no serious opposition beyond the occasional critical remark of a judge or the misgivings of a medico-legal expert.[17] The issue developed differently, however, in the United States. As part of a more general legal reform, the state of New York in 1828 was the first to enact a statute against disclosure of confidential patient information in court: "No person duly authorized to practice physic [that is, medicine] or surgery, shall be allowed to disclose any information which he may have acquired in attending any patient in a professional character, and which information was necessary to enable him to prescribe for such a patient as a physician or to do any act for him as a surgeon."[18]

While the exact circumstances of the introduction of this statute are not known, Justice Buller's statement in the case of *Wilson v. Rastall* and a wish to grant the medical profession the same privilege as the legal profession in keeping communications with clients confidential seem to have been relevant.[19] The commissioners responsible for the revision of the New York statutes gave two reasons for the rule. First, they argued that in comparison with the established privilege for communications between attorney and client, which enabled proper preparation for legal proceedings, consultations with a medical adviser were even more deserving of protection against disclosure. Without it, people would refrain from seeking the medical help they needed. Second, driven by a sense of professional honor, medical men might be tempted to conceal the truth if they were compelled to give evidence about confidential patient details.[20] The position of the Medical Society of the State of New York on this issue probably had played a role here. In its code of conduct, titled *System of Ethics* (1823), the society had declared that it was "a matter of justice, necessity and propriety" that the business of physicians and surgeons should always be considered as confidential and that medical secrecy should be "inviolable even in a court of justice." Comparing the duty of medical confidentiality with the secret nature of the Catholic confessional, the society required doctors appearing as expert witnesses in court to remain silent about matters such as questionable pregnancy and paternity, venereal diseases, alleged disabilities, virginity, and other circumstances that were linked with "a degree of shame" and "never mentioned but with an engagement to secrecy."[21] The 1828 New York statute permitted physicians and surgeons to adhere to this code.

Missouri, in 1835, was the next state to adopt a medical privilege statute.[22] A wave of such enactments followed in the late nineteenth century. By 1889, twenty American states or territories had introduced statutes restricting or prohibiting disclosure of patients' details in court by their physicians or surgeons, unless the patient had consented to it or the duty of medical confidentiality had explicitly been waived. Often, the rules for a medical privilege were set alongside those that protected the communications between attorneys and their clients and the confessions made to clergymen or priests.[23] By September 1895, the number of states or territories recognizing a medical privilege had further increased to twenty-five.[24] The other states continued to follow the English common-law rule that there were no restrictions on disclosure of patient details in court.[25] By the beginning of World War I, Arizona, West Virginia, and the District of Columbia had joined the list of states that had enacted a medical privilege, but initiatives to this effect in Illinois and Massachusetts had failed. Twenty-one American states had no such law in 1914.[26]

The formulations of the relevant statutes varied from state to state, leaving room for different interpretations and applications. For example, the scope of information that was necessary for a doctor to treat a patient, and that was thus protected, was controversial. Moreover, physicians or surgeons consulted for the means of procuring an abortion could, in some states, be forced to testify about this, regardless of a general medical privilege in court. Abortion, then regarded as a crime, was not meant to be "shielded" by medical confidentiality.[27] In several states, the statutory medical privilege applied only to civil actions.[28] In Ohio, it was legal practice to remove the medical privilege in criminal cases—for example, in a trial for rape.[29] Finally, if a patient sought damages from her physician for malpractice,[30] or from another party for personal injury, and for this purpose revealed in court full details of her condition and treatment, she could then lose the right to insist on the physician's confidentiality and to prevent him, or another physician who had been called in as a consultant at this occasion, from giving evidence.[31] This so-called implied waiver of the privilege could also extend to the patient's records in a hospital, as later decided in a personal injury case in Missouri.[32]

The range of medical practitioners to whom the privilege was applicable also was not entirely clear. In 1895, the Supreme Court of Michigan held, against previous opinion, that dentists were not included under the category of "surgeon" in the statutory privilege. It also took the view that the original purpose of the medical privilege was to "invite confidence in respect to ailments of a secret nature" but would not apply to cases "where the infirmity was apparent to every one on inspection."[33] On the other hand, in a personal injury case in 1910, the Supreme Court of Indiana excluded the offered evidence of a medically unqualified superintendent of a gymnasium and orthopedic institute as privileged, although the relevant statute of this state referred to "physicians."[34] And the Supreme Court of Idaho confirmed in an injury case in 1918 that a radiologist's observations were privileged alongside those of the attending physician.[35]

Incorporated into the Code of Remedial Justice (1876) and its successor, the Code of Civil Procedure (1877), the New York statute itself was further revised several times in the late nineteenth and early twentieth centuries.[36] The revisions provided the option for patients or their attorneys to issue waivers of confidentiality in trials and permitted physicians and surgeons to give evidence on the previous mental or physical state of a deceased patient with the consent of a relevant relative or other interested party,[37] as long as the information did not include confidential communications or facts that might disgrace the memory of the patient.[38] In 1904, professional or registered nurses were included under the medical privilege in court, and in 1905, an addition was made to section 834 of the Code of Civil Procedure compelling disclosure of information when a crime against a child under the age of sixteen was suspected.[39]

Medical Confidentiality in German Law

In Germany, yet another approach was taken toward the question of medical secrecy. Here, state involvement in matters of health care was part of the Enlightenment concept of "medical police," which assumed that public measures to control and enhance the health of the population would strengthen a state's political and military power.[40] Within this broader context, medical secrecy was an element of official health regulation. For example, a Prussian regulation forbidding medical personnel to disclose private details of their patients has existed since the early eighteenth century: the Prussian Medical Edict of 1725 ruled that "medical men must not reveal to anyone the secret faults and ailments which they have discovered."[41] A more comprehensive regulation was included in the *Preußisches Allgemeines Landrecht* (Prussian General Law) of 1794, determining that doctors, surgeons, and midwives must not reveal "ailments and family secrets that come to their knowledge, as long as these are not crimes." Offences against this rule carried a fine between five and fifty thalers.[42] Considering that according to the Prussian regulations on medical fees (*Medizinaltaxe*, 1725 and 1815), doctors could charge one thaler for a patient visit, this was a notable but not excessive punishment.[43]

Subsequent legislation was influenced by the French *Code pénal* (1810), which in its article 378 required secrecy of doctors, surgeons, pharmacists, midwives, and others who obtained confidential information through their profession, unless disclosure was demanded by law. Punishment for violations of this article could range from one to six months' imprisonment or fines between one hundred and five hundred francs.[44] Section 155 of the Prussian Penal Code of 1851 similarly ruled that "medical persons and their helpers" and others were punishable with imprisonment up to three months or a fine up to five hundred thalers if they disclosed "without authorization" secrets that had been entrusted to them due to their office, profession, or trade.[45] Some other German states—for example, Hanover (1840), Hessen (1841), and Nassau (1849)—punished breaches of confidentiality only if they had been made with malicious intent or in order to gain unlawful advantages.[46] However, after unification of the German states, the new Reich Penal Code of 1871 followed the Prussian model in its section 300. Including the legal as well as the health professions, it ruled that "lawyers, advocates, notaries, counsels for the defense, physicians, surgeons, midwives, apothecaries, as well as the assistants of these persons are punished with a fine of up to 1,500 marks or imprisonment up to three months, if they reveal without authorization private secrets which have been entrusted to them due to their office, profession or trade."[47] The maximum fine of 1,500 marks would have been a significant deterrent, as it would have amounted to a significant proportion of a doctor's annual income. According to statistics on the annual income of doctors in Hamburg in 1886, for example, about 40 percent of them earned fewer than 5,000 marks, and about 60 percent earned more than 5,000 marks. Only 25 percent had an income of more than 10,000 marks.[48]

Thus, in Germany, the confidentiality of professionals was a general legal duty. Breaches of professional secrecy were punishable under the rules of the penal code unless disclosure had been authorized by the entrusting patient or client or if it was required by law. Legal exceptions to the duty of secrecy pertained to knowledge about plans for serious crimes (section 139 of the penal code), including treason, counterfeiting, murder, robbery, and abduction, as well as a possible bomb attack (section 13 of the Explosives Law of 1884).[49] Reporting to the police or the warning of relevant persons was meant to prevent these crimes. Doctors were also obliged to notify the health authorities of specific infectious diseases under section 2 of the Law on the Combating of Diseases Constituting a Public Danger of 1900.[50] In addition, they had to provide official lists of vaccinated persons under the law on compulsory vaccination against smallpox (1874).[51] Finally, doctors had to report births (if they had been present at them) to the registrar, and directors of mental asylums had to notify the authorities of admitted patients.[52]

Initially, it was unclear whether doctors were entitled to refuse giving evidence in court on the grounds of section 300. However, with the Codes of Criminal Procedure (section 52) and Civil Procedure (section 348) of 1877, which came into effect in 1879, German doctors became entitled to refuse testimony regarding private details of their patients—unless the patient concerned had waived medical confidentiality, in which case the doctor had to testify.[53] Giving doctors an "entitlement" to refuse giving evidence in court stemmed from a line of liberal thought in the early nineteenth century that had already exempted defense counsels, attorneys, and clergymen from the duty of testimony. The legislators believed that these professional groups—and then, in the 1870s, also physicians and surgeons—could be trusted to decide responsibly when testimony without the client's permission might be justified. Also, letting the judge decide on this question was seen as problematic, because he might first need to interrogate the professional concerned in order to arrive at this decision—a procedure that would defeat the object of professional secrecy.[54] Furthermore, section 300 did not apply to legal cases in which a doctor faced charges of malpractice or was accused of demanding excessive fees. Here, the doctor could reveal, without the patient's consent, details of the treatment that were necessary for his defense or to justify his claims.[55]

The rationales behind the German legislation on medical secrecy were reflected by the different headings under which the relevant sections were listed. The regulation of the Prussian General Law of 1794 came under the title "On the Crimes of those who, without being Officials, are especially obliged to the Community," indicating a general, civil-servant-like fiduciary duty to the state rather than recognition of the individual patient's interest in secrecy. In the Prussian Penal Code of 1851, the rule against the breach of confidentiality appeared under the heading "Violations of Honor," which seemed to reflect a shift toward protection of another's personality. Finally, section 300 of the Reich Penal Code of 1871 was listed under the heading "Punishable Self-Interest and Violation of Others' Secrets," recognizing the private sphere of the individual patient.[56] Rationales thus changed from a doctor-centered type of confidentiality to a consideration of patients' privacy. This latter perspective was also reflected in the comments of legal scholars on the "legal good" (*Rechtsgut*) that was protected by section 300. For example, the Halle Professor of Law Franz von Liszt (1851–1919) saw this good in the "interest in the protection of personal and family life against unauthorized intrusion."[57] A decision of the German Supreme Court (*Reichsgericht*) in 1885 made it clear that section 300 protected a person's *interest* in keeping their private secrets confidential.[58] The German legal regulations therefore not only protected medical confidentiality as being instrumental to the provision of health care by supporting a fiduciary and efficient relationship between doctors and patients; they also recognized the intrinsic right of patients to keep their illnesses hidden from others as part of their entitlement to an undisturbed private life.

Comparison

If one compares the outlined legal preconditions for medical confidentiality in court in Britain, the United States, and Germany in the nineteenth century, it appears that Germany, with its recognition of a patient's interest in secrecy, provided the strongest protections, and Britain, with its precedents of rejecting a medical privilege, the weakest. The United States seems to have taken an intermediate position, particularly if one considers that only about half of the states or territories enacted medical privilege statutes and that in some states these statutes pertained to civil actions only. The role of the doctor on the witness stand remained contested. In the following section, I will discuss several legal and medical comments on this issue in the United States.

American Debates on Physicians' Privilege in the Courts

Members of the British legal profession had from time to time expressed their unease about the rejection of a medical privilege in British courts. In addition to Justice Buller's remark of 1792, a statement by the London barrister William Mawdesley Best (c. 1809–69), in his internationally successful handbook *The Principles of the Law of Evidence*, characterized this practice as "a rule harsh in itself, of questionable policy, and at variance with the practice in France, and in some of the United States of America."[59] This opinion contrasted, however, with the views of several American legal experts who criticized the New York statute and its successors.

The New York barrister Charles A. Boston, for example, who had conducted a detailed analysis in 1894 of the various medical privilege statutes and relevant legal cases, arrived at the conclusion that the statutes "have not proved an unalloyed benefit, and some of their features have brought about conditions which in some cases have embarrassed the administration of justice."[60] Concerning the New York statute, he highlighted that it prevented a physician from "disclosing the condition of his patient who is a lunatic or habitual drunkard," as well as from stating a patient's cause of death, and that it excluded much testimony that might have demonstrated fraud in insurance cases.[61] Such unintended consequences meant that the statutory medical privilege could restrict the evidence that physicians and surgeons were permitted to give in court to such an extent that their testimony could become almost meaningless. Moreover, the confidentiality required by the privilege statute was used as an instrument to silence a doctor if, for example, his testimony was likely to contradict the claims of a plaintiff who sought damages for injuries.

Another barrister, William Archer Purrington (1852–1926), who had served as counsel for the New York County and New York State medical societies, reported a case that illustrated this problem. A woman who had fallen in a village street sued the local authorities for damages for her injuries, which were listed as umbilical hernia, prolapse of the uterus, and diverse bruises. During the trial, the defense for the village tried to show, by calling the plaintiff's physician, that she had already had the hernia before the accident. The physician testified that what he knew about her health came from attending her professionally over a period of eight to ten years, including two births. The defense now argued that if he had discovered the hernia when she was giving birth and had not treated her for it, this information would not be covered by the statutory medical privilege, and he could therefore testify to this point. The court, however, excluded the doctor's testimony on the grounds that he would still have acquired his information in a professional capacity. The plaintiff was eventually awarded $2,500 in damages.[62] Obviously, in this case, it had not been in the patient's interest to waive medical confidentiality.

Purrington claimed that the introduction of the medical privilege in New York and other states had opened the door to fraud. His experience in this regard had been similar to Charles Boston's. In some legal actions brought to recover damages for physical injuries, the statutory privilege had been used to suppress the "best available evidence" (that is, the medical evidence) if this was in the plaintiff's interest. He therefore called for changing, if not altogether repealing, the privilege statute.[63]

Boston's and Purrington's assessments that the medical privilege in court was being misused for fraudulent purposes matched that of Tracy Chatfield Becker (1855–1935), professor of criminal law and medical jurisprudence at the University of Buffalo. In his view, the relevant statutes had originally been established on the basis of "sound public policy" but operated under conditions that often led to "perversions of justice." Speaking to a national assembly of railway surgeons in Chicago in 1895, he called on them to apply the utilitarian principle of the "greatest good to the greatest number" to the question of their privilege in court. While he acknowledged that their knowledge about the nature and cause of an injury might be regarded as confidential, he still asked them to consider that they might be the only truthful witness for a railway company to show that a patient was responsible for an injury herself through her own negligence. He doubted that patients in general concealed information from their doctors because of the possibility that it might be used against them in court. In other words, he suggested that the surgeons should testify out of duty to their employer (i.e., the railway company) and in line with public sentiment and that the medical privilege should be modified accordingly.[64]

Following a similar line of argument, in 1905, John Henry Wigmore (1863–1943), professor of the law of evidence at Northwestern University, asked four critical general questions about a privilege for the communications between doctor and patient: "[1] Does the communication originate in a confidence? [2] Is the inviolability of that confidence vital to the due attainment of the purposes of the relation of physician and patient? [3] Is the relation one that should be fostered? [4] Is the expected injury to the relation, through disclosure, greater than the expected benefit to justice?"[65] Wigmore answered only the third question in the affirmative. Regarding the first question, he maintained that patients had a real interest in keeping their condition secret in only a few instances, such as in cases of venereal disease or abortion. Concerning the second question, he argued that the possibility of disclosure in court would not deter people from seeking medical help. Polemically, he asked, "Is it noted in medical chronicles that, after the privilege was established in New York, the floodgates of patronage were let open upon the medical profession, and long concealed ailments were then for the first time brought forth to receive the blessings of cure?"[66] Finally, he answered the fourth question by claiming that "injury to justice by the repression of the facts of corporal injury and disease" was "a hundred fold greater than any injury which might be done by disclosure," particularly in divorce proceedings or in cases of criminal abortion. Therefore, in Wigmore's opinion, demands for a medical privilege were unjustified. More than this, he suggested that the medical profession made such demands only out of jealousy of the legal profession, because the latter had a recognized privilege for attorney-client communications.[67] Like Boston, Purrington, and Becker, he was particularly concerned that the medical privilege was misused for concealing relevant information when patients made financial claims in insurance cases—for example, the actual extent of a corporal injury or, in contested life insurance claims, the true previous state of health of the deceased.[68]

The criticisms by these legal commentators reflected a legal interpretation of the medical privilege that gained ground with the introduction of "waivers" of confidentiality by patients or their representatives in court. While the initial conception of the medical privilege in 1828 had aimed to protect confidentiality in the interests of patients as well as their doctors, the practice of waiving confidentiality meant that, actually, only the patient had the right to decide whether her doctor should give evidence about her condition.[69] An apparent exception was the 1891 injury case of *Treanor v. Manhattan Railway Company*, in which the judge, Justice Pryor, held that it was wrong to exclude the testimony of the plaintiff's physician after she had testified to her injuries in great detail. Her conduct, the judge explained, constituted a waiver of confidentiality, and the physician could thus be called in order to (potentially) contradict the patient.[70] The court of appeals, however, took the view that the Treanor case "went too far" in this regard.[71] Nevertheless, twenty years later, in a personal injury case heard by the Supreme Court of Iowa, Pryor's line of argument was taken as well, and the privilege of the plaintiff's physician was regarded as waived because the patient (plaintiff) had testified at length to the pains she suffered and her treatment by the physician.[72] On the other hand, in the same year, 1911, the

Supreme Court of Arkansas ruled in an injury case that a patient's waiver of secrecy for one particular physician did not entitle the opposite side to call as a witness another physician who had treated the patient on another occasion for the same condition. This decision was based on the view that the statutory privilege had been introduced for the benefit of the patient.[73] And in another personal injury case, in 1913, the Supreme Court of Wisconsin confirmed that the patient had not waived her right to medical confidentiality under the privilege statute by testifying concerning the nature and extent of her injuries.[74] The applicability of the medical privilege was thus contested and interpreted in different ways among the various states that had adopted it. The option for the patient to insist on medical confidentiality created problems for the courts in accessing all relevant evidence, and the notion of "implied waivers" was used by some judges to circumvent this obstacle.

On the other hand, from the side of the medical profession, it was maintained that the "machinery for the conviction and punishment of crime" operated as well in those states that had a medical privilege in court as in those that did not. This point was made by Daniel R. Brower (1839–1909), professor of mental diseases at Rush Medical College, Chicago, and Northwestern University in 1896, just a year after Becker's speech, in an address to the local Medico-Legal Society. Brower argued for the introduction of a statutory medical privilege in the state of Illinois, claiming that his medical colleagues in the society would rather go to prison for contempt of court than violate secrecy and expose the character of their patients in a court of law.[75]

In fact, the medical sentiment that patients' confidences must be strictly protected had been part of the Code of Ethics of the American Medical Association (AMA) since its introduction in 1847. In section 2, the code stated, "Secrecy and delicacy, when required by peculiar circumstances, should be strictly observed; and the familiar and confidential intercourse to which physicians are admitted in their professional visits, should be used with discretion, and with the most scrupulous regard to fidelity and honor."[76] While this first sentence had been adopted almost verbatim from Percival's *Medical*

Ethics,[77] the section continued, "The obligation of secrecy extends beyond the period of professional services—none of the privacies of personal and domestic life, no infirmity of disposition or flaw of character observed during professional attendance, should ever be divulged by him [i.e., the physician] except when he is imperatively required to do so. The force and necessity of this obligation are indeed so great, that professional men have, under certain circumstances, been protected in their observance of secrecy by courts of justice."[78] When a revised version of the code was proposed by an AMA committee in 1894, another sentence was added to this section: "Physicians should discourage as much as possible, however, the needless disclosure of family concerns, the knowledge of which is not essential to successful treatment."[79]

The AMA code obviously reflected the challenges to confidentiality that doctors had experienced in court and the protection of patient information provided through the 1828 New York statute. The additional sentence from 1894 may have been suggested in view of the difficulty in defining the scope of information that was privileged: if the knowledge was not strictly related to medical or surgical treatment, its disclosure might be compelled in court. However, the AMA's national meeting of 1894 overwhelmingly rejected the proposed revised code because it included far-reaching reforms of other, highly contentious subjects, such as legitimization of advertising by specialists and patenting of instruments by surgeons and acceptance of homeopathic and female physicians as peers.[80]

Neither was the cited last sentence included in the code's successor, the "Principles of Medical Ethics," which were adopted by the AMA in 1903. While the "Principles" marked a transition in AMA policy toward less stringent guidance, leaving ethical decision making largely to the local medical societies,[81] the sections on secrecy were not only preserved but clarified in an important point: disclosure of confidential patient information was only permissible "when imperatively required by the laws of the state."[82] In other words, only explicit legal requirements could condone a breach of medical confidentiality.

Brower's expectation in 1896 that his colleagues would rather go to prison than betray patients' confidences had a prominent predecessor in a statement by Austin Flint (1812–86), one of the most renowned American clinicians of the time, who had been president of the New York Academy of Medicine and the AMA. In his book *Medical Ethics and Etiquette* (1893; first pub. 1882)—a detailed commentary on, and in defense of, the AMA's Code of Ethics[83]—Flint discussed, among other things, the dilemma in court for those doctors who lived in a state that had not recognized a medical privilege. When required to disclose confidential information about a patient, either they might follow their ethical convictions and refuse to testify, risking punishment for contempt of court, or they might follow the law and give evidence, acting in opposition to their consciences and medical ethics. Ideally, Flint said they should seek to secure in their state an enactment (like in New York) that removed the conflict between the Code of Ethics and the law.[84] In the meantime, however, Flint expected his colleagues to refuse to testify: "Penalties incident to the conviction that the requirements of professional confidence should be inviolable are to be accepted as conducive to the honor of the profession and to humanity."[85] In a similar vein, the president of the Medical Society of the District of Columbia, Samuel C. Busey, praised in his annual address of 1899 the adoption of privilege statutes as adding "the force and fiat to the decree of the medical profession, which has always and everywhere throughout the civilized world resisted the compulsory disclosure in open court."[86]

Yet the opposition of some legal professionals to a medical privilege made it difficult for physicians to achieve legislation on this matter, as is illustrated by the case of Illinois. In January 1897, physician F. L. Hall of Perry (Illinois), a member of the legislature, introduced a bill providing for medical secrecy along the lines of the New York statute.[87] By the end of May, his colleague, John B. Hamilton, professor of surgery at Rush Medical College, had to report that Hall's bill had already failed at the committee stage. With bitter irony, Hamilton described how "an old lawyer with snowy looks and hands trembling with age wrought himself into a storm of passion in denouncing it" and how the "awe-struck committee, conscience-smitten for having for a moment dared to look at a bill introduced by a physician," rejected the proposal.[88] The legal view that a medical privilege would pose an obstacle to the course of justice by excluding important evidence had obviously prevailed. At about the same time, physician James B. Baird campaigned in Georgia for the enactment of a medical privilege statute. Emphasizing the fact that twenty American states had such regulations by then, he claimed that public opinion was "decidedly in favor of protecting the professional secrets of physicians."[89]

Commentators thus expressed widely different views on what the public thought about the need for protecting medical confidentiality in court. Baird's assessment of public opinion on this question contrasted with that of Becker and Wigmore, who held that most people did not worry about this matter. Nevertheless, the intensity of the medico-legal debate in the 1890s about the implications and consequences of medical privilege statutes suggests that it may have been fueled by more general concerns about privacy—at least among the social and intellectual elite. In December 1890, the now-famous article "The Right to Privacy" by Boston attorneys Samuel Warren (1852–1910) and Louis Brandeis (1856–1941) had been published in the *Harvard Law Review*.[90] Drawing upon English legal cases, they argued for recognition of an individual's "right to be let alone."[91] It has been said that the immediate occasion for writing on privacy was Warren's annoyance about newspaper reports on his and his wife's glamorous dinner parties, and the main thrust of the article aimed at the practices of the press, including the use of photographs made with the then newly available "snap cameras."[92] Warren and Brandeis described, however, a much more general need for privacy in modern, complex society and recognized that invasion of an individual's privacy caused "mental pain and distress, far greater than could be inflicted by mere bodily injury."[93] While they did not directly discuss the disclosure of patients' secrets,[94] their general assessment of people's need for "solitude and privacy" indicated a contemporary social climate that was conducive to protecting medical secrecy. However, the lines between certain members of the legal profession who criticized a medical privilege in court and members of the medical profession who defended it (or campaigned for it) seem to have been fairly clearly drawn in the United States.[95]

Medical Privilege in Germany

In the immediate years after the regulation of professional secrecy through section 300 of the Reich Penal Code, doctors' communications with their patients were still unprotected in German courts. While lawyers and priests were not required to testify, physicians and surgeons could be forced to give evidence about their patients. In 1875, the Reich Justice Commission (*Reichsjustizkommission*) planned a right for courts to compel any expert to testify, because the expert "owed his knowledge to the state." From the perspective of the medical profession, this would have been an unjustified expropriation of intellectual property.[96] Eventually, from 1879 onward, section 52 of the Code of Criminal Procedure and section 348 of the Code of Civil Procedure granted doctors the "entitlement" to refuse giving evidence in court on the grounds of their legal duty of professional secrecy according to section 300 of the Penal Code.[97]

However, this situation raised the question of whether doctors were obliged to make use of this entitlement or whether they could choose to testify without their patient's consent. District court judge Liebmann, author of a book on medical secrecy in 1886, held that doctors did risk punishment under section 300 if they gave evidence without permission of the patient—a view still shared in 1904 by a senior judge, *Oberlandesgerichtsrat* Simonson in Breslau.[98] Other legal commentators took the view, however, that giving evidence in court, at the request of a judge, could never be an illegal act.[99] The German Supreme Court (*Reichsgericht*) confirmed in a decision in 1889 that a doctor was free to decide in each particular case, according to his own "dutiful judgment and discretion," whether he wanted to testify concerning his patient.[100] Still, as long as there had not been a plenary decision of the Supreme Court on this question, doctors who did not want to take any risk of punishment for unauthorized breach of secrecy were advised to refuse to give evidence in any case, unless the patient concerned had explicitly released them from the duty of confidentiality.[101]

Moreover, in German legal practice, the entitlement of doctors to refuse to give evidence in court was not accepted without challenges. For example, in 1884, the District Court of Frankfurt/Main had to confirm through an interim judgment in a divorce case that the attending physician and the hospital's administrator were entitled to refuse testimony regarding the alleged treatment of the husband for venereal disease.[102] More than twenty years after the Code of Civil Procedure had come into force, a Hamburg physician was involved as a witness in a protracted divorce case that led to two decisions of the German Supreme Court regarding his medical privilege. After the doctor had refused to testify to the husband's disease on the grounds of section 383 (formerly section 348) of the Code of Civil Procedure, section 300 of the Penal Code, and his personal promise of confidentiality to the patient, the Higher District Code of Hamburg ordered the husband in an interim judgment in June 1901 to release his doctor from the duty of confidentiality so that the latter could confirm in court that he had treated him for freshly acquired syphilis in May 1899. With such a confirmation, the wife intended to provide the required proof of the husband's adultery and to obtain the desired divorce. The doctor, however, continued to refuse his testimony. The Higher District Court convicted him in March 1902 and ordered him to give evidence. His subsequent complaint to the Supreme Court was found to be justified, the judgment against the doctor was quashed in April 1902, and the case returned to the lower court.[103] Yet, in September of the same year, the Higher District Court again asked the doctor to testify, indicating that his testimony would not be punishable under section 300 of the Penal Code. He refused again, emphasizing that he had personally promised the husband to keep confidentiality.[104] After yet another conviction of the doctor by the Higher District Court and yet another complaint by him, the Supreme Court eventually ruled, in January 1903, that the doctor was indeed entitled to refuse his testimony regarding the husband's disease. The legal costs for the interim proceedings had to be paid by the wife as the plaintiff.[105]

The Hamburg case illustrated how the implementation of the legal entitlement of German doctors to refuse testimony could be controversial, but this entitlement was eventually assured by the highest court, at least for civil proceedings. In its reasons for the verdict, however, the Supreme Court admitted that there might be cases in which "higher moral duties" could overrule the duty of confidentiality. As an example, the court suggested that a doctor might feel obliged to inform a wife of the venereal disease of her husband in order to protect her against infection as far as possible. Going a step further, the court did not even rule out a moral duty to inform a third party other than the wife. In this particular case, however, the testimony sought from the doctor, concerning the husband's syphilis infection some years ago, was meant not to maintain the wife's physical health but to support her wish for a divorce. Compared to the duty of secrecy, in the court's opinion, facilitating a desired divorce could not be regarded as the *higher* moral duty. The doctor's refusal to give evidence was therefore seen as justified.[106]

In a roughly parallel divorce case, at the District Court of Mainz, a doctor likewise refused to testify to a wife's claim that her husband had infected her with syphilis and that consequently their three-year-old child had also acquired the disease. The couple and their child had been his patients. While the wife had waived confidentiality, the husband explicitly declared that he did not release the doctor from his duty of secrecy. In this case, the court ruled that the doctor had to give evidence of the wife's disease but not that of the husband or, due to paternal authority, the minor child. The significance of this case was that the court separated out the individuals entrusting a secret to the family doctor and made him testify despite his perception of a joint family secret, depending on who had released him from the duty of confidentiality.[107]

The complexities of the medical privilege in such cases showed that doctors in Germany could not be certain that courts would always honor their entitlement to decide themselves whether they wanted to testify. Despite the relative clarity of the German law on professional secrecy, medical confidentiality was open to legal challenges—similar to the situation in the United States described above.

The Lack of a Medical Privilege in Britain

Based on the precedent of the Duchess of Kingston case, in Britain, doctors were still compelled to give evidence on patient details in the late nineteenth century. The London obstetrician John Braxton Hicks (1823–97), for example, experienced such a scenario. In a divorce case, he was subpoenaed to testify about his consultation of the husband, who, a year earlier, had come to see Hicks with worries that he might have infected his then-pregnant wife with gonorrhea. Reluctant to speak, Hicks asked the court whether the husband's communication to him might be privileged but was told that it was not, and he subsequently gave evidence. Hicks was particularly annoyed that he was put in a situation in which he was forced to provide a testimony that, in effect, was equivalent to a self-incrimination of one of his clients—that is, the husband.[108]

In the contemporary British medico-legal and ethical literature, doctors were advised to testify only after a ruling of the judge to this effect[109] or even to refuse giving evidence altogether, risking a prison sentence for contempt of court, "which might be the more honourable course."[110] As Justice Sir Henry Hawkins (1817–1907) emphasized during a prominent libel case against the obstetric physician William Smoult Playfair (1835–1903) in 1896 (*Kitson v. Playfair*), the question of privilege was for the judge to decide, depending on the particular circumstances of a case.[111] Usually, this meant that the doctor concerned was compelled to testify. Only in some exceptional circumstances did a judge exempt a doctor from giving evidence, as in a matrimonial case before the Nottingham Magistrates in 1900. Here, the medical evidence might have incriminated the female defendant, and the judge recognized the doctor's concerns that, besides being a breach of the ethical duty of confidentiality, disclosure might make him liable to action by the defendant as well as by his professional body, the General Medical Council.[112]

In fact, in the previous year, in January 1899, the president of the General Medical Council had a memorandum on professional secrecy of medical practitioners prepared by the council's legal assessor, Muir Mackenzie.[113] Its original purpose was to inform the secretary of state, who had received a query on this topic by the Russian ambassador. However, the memorandum was subsequently also published in *The Lancet* and thus brought to the wide attention of the medical profession. Briefly reviewing the relevant legal cases since the trial of the Duchess of Kingston, Mackenzie concluded that "a medical man not only may, but must, if necessary, violate professional confidences when answering questions material to an issue in a court of law."[114] Moreover, drawing upon Justice Hawkins's summing up in the *Kitson v. Playfair* trial, he warned that "circumstances which according to the custom of the medical profession might be deemed to exonerate him [that is, a medical man] from the imputation of improper violation of secrecy might nevertheless in a court of law be deemed an insufficient justification."[115] The circumstances referred to were criminal communications and protection of a doctor's own wife and children.[116] British doctors were thus in a position where they could be forced to give evidence in court but were simultaneously expected to observe strict confidentiality in daily life. A breach of confidentiality could result in charges of slander or libel and might end with a verdict to pay considerable damages to the patient, as had happened in the *Kitson v. Playfair* case.[117] As the London ophthalmic surgeon and General Medical Council member Robert Brudenell Carter (1828–1918) pointed out in 1903, the absence of a privilege for medical witnesses was "now the unquestioned law of the land," but apart from this, a doctor making "any indiscreet disclosure by which the patient sustained injury, would be liable to be cast in damages."[118]

The issue of a medical privilege in court became prominent again in the years after the end of World War I, when the number of divorce petitions soared. Typically, medical evidence was sought in such cases to prove that the husband had acquired a venereal disease outside marriage and subsequently infected the wife. In this way, the wife could provide the required proof of adultery and cruelty. With the Public Health (Venereal Disease) Regulations of 1916, the English government had established special VD treatment centers that guaranteed confidentiality. This circumstance made the requirement of disclosure in court especially problematic for those doctors who worked at these centers and were subpoenaed to give evidence in divorce trials. Two such legal cases, *Garner v. Garner* (1920) and *Needham v. Needham* (1921), attracted much attention.[119] In the first case, the doctor, Salomon Kadinski of Westminster Hospital, protested against the request to testify on behalf of the wife, who claimed to have been infected by her adulterous husband. The judge, Justice Henry Alfred McCardie (1869–1933), did not recognize the protest, stating that there were "even higher considerations" in a court of law than those pertaining to the position of medical men. Kadinski subsequently testified, with the wife's consent, that she suffered from syphilis.[120]

In the second case, Dr. John Elliott (1861–1921), superintendent of the VD clinic in Chester, had been subpoenaed to testify to the husband's claim that his wife had contracted gonorrhea in an adulterous relationship. Encouraged by the newly established Ministry of Health, which hoped to make this a test case for the confidentiality of the VD treatment scheme and a medical privilege in court in this respect, Elliott initially refused to give evidence, referring to the authority of the 1916 regulations. However, the judge, Justice Horridge, did not regard this authority as sufficient to justify a medical privilege in court and ordered him to testify. Faced with the prospect of a six-month prison sentence for contempt of court, Elliott acquiesced and gave evidence.[121]

Attempts by the Ministry of Health to secure a medical privilege for civil proceedings failed due to the resistance of the judiciary, led by the lord chancellor, Viscount Birkenhead (F. E. Smith, 1872–1930).[122] The influence of the British Medical Association in the matter was hampered by internal differences. On the one hand, the delegates of its annual representatives meeting demanded in 1921 that doctors in a situation like Elliott's should be guaranteed the association's support. On the other hand, the BMA's council and central ethical committee largely defended the status quo on the issue of medical confidentiality in court.[123] In 1922, Birkenhead published a strong defense of the traditional view that doctors had no privilege in court and had to support the administration of justice. In his opinion, "to establish a class who may at their will assist or obstruct the judges in their work would be a retrograde step not justified by any argument which has been brought forward."[124] In a 1926 textbook on medical law, London barristers William Sanderson and E. B. A. Rayner confirmed the lack of a privilege in court for the medical profession. They acknowledged that medical confidentiality helped efficient treatment by encouraging patients to speak openly with their doctors about their health problems. However, knowing that what they said to their doctor might later be used as evidence against them would make some patients reluctant to disclose matters that might be important for their health. But as Sanderson and Rayner pointed out, the law on privilege remained valid regardless of such considerations: "The real basis of the claim of the medical profession for privilege [. . .] rests upon the maintenance of the efficiency of the function of medicine and its service to the body politic. This claim necessarily gives rise to differences, which at present appear to be irreconcilable."[125]

Indeed, in another divorce trial, in 1927 in Birmingham, Justice McCardie again demonstrated his uncompromising legal position: he compelled medical evidence on the husband's alleged venereal disease, regardless of the governmental guarantee of confidentiality in VD treatment centers and against the protest of the medical staff concerned.[126] Speaking subsequently to the Medico-Legal Society, McCardie further defended his position. As the London correspondent of the *Journal of the American Medical Association* reported the judge's speech: "There were two aspects of the question [of medical secrecy], each of which was vital. There was the physician who said, 'Health, health, health, and break down the legal obstacles that prevent the gain of health.' Yes, but there was another point of view—

and there was not a lawyer whose heart was not stirred—and that was 'Truth, truth, truth; open the shutters and let in the full light of truth.' Truth lay at the root of criminal justice."[127]

A private member's bill to allow a medical privilege regarding VD cases, introduced into Parliament in 1927 by the dermatologist and MP for the University of London, Ernest Gordon Graham-Little (1867–1950), was unsuccessful, as was his second attempt in 1936–37 with a bill for a wider medical privilege. Lacking ministerial support and unclear in its potential consequences, Graham-Little's initiative faltered and failed under the pressure of legal criticisms.[128] Especially the case of *Garner v. Garner* had set an important new precedent for the lack of a medical privilege in British courts. It is still cited nowadays on this point.[129]

Conclusions

As this chapter has shown, the course of policies regarding medical confidentiality in court in Britain, the United States, and Germany was determined by decisions or regulations that reached back to the eighteenth and early nineteenth centuries. In all three countries, the question of a medical privilege was to some extent contested, but the outcomes established by the early twentieth century differed considerably. Although British medical practitioners, when called as witnesses, repeatedly tried to maintain secrecy about private patient details, judges compelled them to testify on the basis of legal precedent and the view that the medical evidence was material to the case concerned. In the United States, by the end of the nineteenth century, about half of the states had followed the example of New York and had adopted statutes that restricted disclosure of patients' information in court by their physicians or surgeons, but exceptions were widely recognized. This was especially true for criminal cases, such as illegal abortion, where the ethical duty of confidentiality was overridden by the interest in prosecution and the securing of convictions. In civil actions, the legal profession was considerably concerned that the medical privilege was abused to commit fraud, especially in personal injury and life insurance cases. In Germany, doctors' legally guaranteed entitlement, from the late 1870s onward, to refuse to give evidence in criminal and civil cases did not prevent serious legal challenges that went right up to the Supreme Court. Eventually, however, medical secrecy was protected by the German courts on the basis of section 300 of the Reich Penal Code, unless there was a "higher moral duty" that might justify disclosure—for example, warning the wife of a syphilitic husband of the danger of infection.

In part, these differences had to do with differing legal frameworks. In Britain, English common law's reliance on precedents posed an unsurmountable obstacle for recognizing medical confidentiality in court since the Duchess of Kingston case of 1776.[130] In Germany, by contrast, the confidentiality of medical practitioners had been made a requirement in statute law since the eighteenth century as an element of state health policy. In the American states, the situation for doctors called to give testimony varied depending on whether their state continued to follow English common law or whether it had adopted a form of medical privilege following the example of New York in 1828.

The different situations for doctors in British, American, and German courts also appear to have resulted from different power relations between the medical and legal professions. Whereas in Britain the judiciary's interest in unrestricted access to medical evidence dominated over doctors' attempts to protect confidentiality (as illustrated especially by the rulings of Justices McCardie and Horridge), a more balanced relationship between the two professions in Germany led to recognition and confirmation of a medical privilege in court. When German doctors were granted the legal right to refuse to give evidence in court in the late 1870s, one important consideration had been that they could be trusted, in the same way as the members of the legal profession and the clergy, to exercise discretion and to decide when to stay silent and when to testify, even against the patient's wishes. This indicates a relatively high social status of German doctors in the late nineteenth century, at least in comparison with their British colleagues. The United States, with about half of them adopting a medical privilege and the other half still adhering to the English common-law rule, seemed to reflect, apart from differences in the influence of the local medical profession, the outcome of efforts of traditionalist forces in law to withstand the modernizing model of the New York statute. The failed attempt to introduce a medical privilege statute in Illinois in the late 1890s points to this interpretation. Even in those American states that had enacted a privilege for communications with physicians and surgeons, there was still considerable uncertainty about the specific circumstances in legal cases to which it applied or in which it might be regarded as implicitly waived.[131] The notion of the "implied waiver" of confidentiality, if the patient had herself revealed details of her condition in court, was an instrument for American judges to invalidate a patient's wish to prevent her doctor from giving evidence despite an existing statutory medical privilege.

Generally, secrecy was an asset for the medical profession that set it apart from unlicensed practitioners. The prominence given to medical secrecy as the second section of the American Medical Association's Code of Ethics reflects this status. Besides confidentiality's importance for maintaining the trust between doctor and patient, a right to remain silent in court reflected a claim to a status that equaled that of the legal profession. Toward the end of the nineteenth century, recognition of an individual's right to privacy began to provide an additional argument for protecting the medical secret against disclosure. In German law, as von Liszt explained, medical confidentiality protected people against intrusion into their personal and family life, and it recognized their interest in keeping their illnesses secret (unless one of the specific legal exceptions, such as prevention of serious crimes or of dangerous contagious diseases, applied). In America, attorneys Warren and Brandeis had laid the legal groundwork for recognizing a general right to privacy. In Britain, however, confidentiality, while demanded in works on medical ethics since Gregory and Percival, had experienced its limits in the courts of law. The next chapter will address a specific point of conflict between those who wished to protect medical confidentiality and those who believed that disclosures—not just in court but generally—were justified in the public interest: the case of venereal diseases.

Chapter 2

Venereal Diseases: The Issue of Private versus Public Interest

Introduction

In 1920, the Supreme Court of Nebraska confirmed the "not guilty" verdict for a physician who had been sued by a patient because of breach of medical confidentiality. The court believed that this case was a "novel one," the first of its kind being heard at a court of final appeal. The details of the case, which became known in the literature as *Simonsen v. Swenson*, are as follows.

Working for a telephone company in another town, Mr. Simonsen was temporarily staying at a small, family-run hotel when he noticed sores on his body and consulted a local physician, Dr. Swenson, who acted as the hotel doctor when required. After examining the patient, the physician diagnosed syphilis, though, as he admitted, he could not be entirely sure about this without a Wassermann test,[1] for which he had no equipment. He asked Mr. Simonsen to leave the hotel the next day because of the danger of infection for other guests, and the patient promised to do so. The next day, Dr. Swenson, who was also the family doctor of the hotel manager, visited the latter because of illness and learned at this occasion that Mr. Simonsen had not left the hotel. The physician warned the hotel manager's wife that this guest suffered from a "contagious disease" and advised her to be careful, to disinfect his bedding, and to wash her hands in alcohol afterward. She subsequently put all of Mr. Simonsen's belongings in the hallway, fumigated the room, and forced him to leave the hotel. Having left, Mr. Simonsen had a Wassermann test done by another doctor in another town. The test result was negative, but the doctor was unsure whether the disease might nonetheless be syphilis.[2]

At the time, Nebraska had a medical privilege statute (section 7898 of the Revised Statutes of 1913), but obviously, since the disclosure of the patient's illness had happened outside of court, this was not applicable to this case. The plaintiff was able to refer, however, to section 2721 of the Nebraska statutes, which said that a physician's license could be revoked because of "unprofessional or dishonorable conduct," including "betrayal of a professional secret to the detriment of a patient." Nevertheless, the court found that the defendant had been justified in warning the hotel manager's wife of her guest's infectious disease. Given its highly contagious nature, the physician was entitled to "make so much of a disclosure to such persons as is reasonable and necessary to prevent the spread of the disease."[3] The court also considered that, under Nebraska statutes, the state board of health and municipalities required reports of dangerous infectious diseases and provided quarantine rules.[4]

The case was subsequently discussed in the legal as well as the medical press.[5] As a commentator in the *Harvard Law Review* put it, "The case stands for the triumph of medical altruism over legal duty. It sanctions the assumption by the doctor of the police power of the state with regard to disease in contravention of his relational duty to his patient, and grants the obligor the discretion to perform his duty or not."[6] Another commentator, in the *Yale Law Journal*, concluded from the outcome of the case that "to preserve silence is a duty owed to the individual, but only as a member of society; and when absolute silence becomes detrimental to the public welfare, the duty ceases."[7] And an editorial in the *Journal of the American Medical Association* even claimed that the case established for the first time through a Supreme Court judgment "the exact relations that exist between a physician and his patients, on the one hand, and a physician and the public, on the other."[8]

The case of *Simonsen v. Swenson* touched on several issues that were characteristic of the contemporary discussions on confidentiality in venereal diseases. It was controversial whether an individual patient's interest in keeping their condition secret should be overridden by the public's interest in being protected against infection and whether it was therefore permissible to breach the patient's confidence in order to warn potential contacts. Furthermore, it was undecided whether venereal diseases, regardless of the social stigma they carried, should be reported to the health authorities like other contagious diseases, such as smallpox, cholera, or diphtheria. There were concerns that the prospect of notification might deter VD-infected persons from consulting a doctor and that the reporting of cases of venereal disease might have a detrimental effect on a doctor's practice or the reputation of a clinic. These and related issues were at the heart of the debates on VD and confidentiality in the late nineteenth and early twentieth centuries. In this chapter, I will explore how medical and legal professionals in the United States, Britain, and Germany responded to these issues.

Venereal Disease Notification in the United States

The case of Dr. Swenson may have attracted attention not only because a patient had sued a doctor for breach of secrecy but also because the doctor's conduct was unusual for this time. Due to the social stigma attached to venereal diseases, physicians were typically reluctant to disclose information on patients with these conditions, even if there was a legal requirement to notify the health authorities. A key argument was that the confidentiality of the doctor–patient relationship had to be preserved in order to secure patients' cooperation for effective, continuous treatment. A syphilis infection then commanded extensive therapy with mercury preparations, iodides of potassium, or, from 1910, the new arsenical compound Salvarsan. Gonorrhea was treated with various "antiseptics," which were given orally or directly injected into the urethra.[9] The *Simonsen v. Swenson* case came in the midst of debates between public health officers and private physicians about the necessity of reporting VD cases (that is, syphilis, gonorrhea, and chancroid) to the health authorities in order to aid disease surveillance and, if necessary, intervention. From the early twentieth century, state and municipal boards of health had tried to introduce mandatory reporting of VD cases. Widespread concerns over the health of soldiers and of the workforce at home during the First World War supported these efforts, so by 1919, all American states required some form of notification of venereal diseases. In most states, anonymous reporting (by a code number or use of the patient's initials only) was permissible and practiced, although public health officials tended to favor full notification including the patient's name and address.[10]

However, compliance of physicians was hard to achieve.[11] Just a few months before the case of *Simonsen v. Swenson* became widely known, William Edler, the director of the Bureau of Venereal Diseases of the Louisiana State Board of Health, had given a polemical address at the 1920 meeting of the American Medical Association in New Orleans. Exasperated by physicians' unwillingness to report, a problem that had previously been encountered with tuberculosis notification,[12] he suggested that doctors who did not comply with the notification laws for venereal diseases should be punished with temporary withdrawal of their practicing license: "One or two prosecutions of this kind in a community, with ample publicity, will produce more communicable disease reports than will years of propaganda."[13]

Edler identified three types of objection against VD reporting, all of which, he claimed, were invalid: first, that it violated the patient's right to keep his disease secret as well as the established principle of privileged communications between physicians and their patients; second, that it would expose doctors to serious risks, such as blackmail, prosecution for libel, charges of malpractice, or even "being shot"; and third, that in case of married male patients, if their wives found out, "domestic turmoil and tragedy" (that is, divorce) would follow. For Edler, the first objection had no force because individual rights had to yield to the social interest of disease prevention, and state-licensed doctors had, in his opinion, a duty to protect the public against infection. In view of the second objection, he pointed out that Ohio had the second highest number of VD reports in the United States, although it operated a notification system by name and address, and none of the feared incidents had happened. Surely, then, doctors in states that only required reporting with anonymous code numbers had no reason to worry. Finally, regarding the potential impact on family life, Edler rhetorically asked his (male) medical listeners whether they wanted to protect their own sex with their sympathy for the VD-infected husband (who had usually acquired the disease from a prostitute or adulterous affair) at the expense of the wife, who would be infected through him, and their children, who would suffer from "hereditary" syphilis.[14]

In fact, as historian Allan Brandt has shown, concerns over the threat posed by VD to the middle-class American family motivated certain Progressive-Era physicians to campaign in a "war" against venereal diseases. Declining birth rates in the white population were attributed to infertility caused by venereal infection, particularly gonorrhea, in women, and so-called hereditary or congenital syphilis raised eugenicists' fears over racial degeneration.[15] However, the fight against venereal disease did create serious ethical conflicts with doctors' duty of confidentiality. One of the most active anti-VD campaigners, New York dermatologist Prince A. Morrow (1846–1913), founder of the American Society of Sanitary and Moral Prophylaxis, devoted a whole chapter of his 1904 book *Social Diseases and Marriage* to the "medical secret" in this context. This allows us to gain more detailed insight into where the perceived ethical problems lay.

Secrecy, Marriage, and VD Prevention in the United States

While Morrow recognized medical secrecy as "the primal professional virtue" and "the basis of all relations between the physician and patient," he also saw it as a major obstacle to the physician's public health duty of preventing the introduction of venereal disease into marriage.[16] An "unfortunately not rare" situation, in Morrow's experience, was when a VD-infected man did not want to desist from his marriage plans, regardless of his physician's warnings and representations about the danger of transmitting the disease to his future wife and their offspring. Here, the physician was caught between his duty to protect the confidence of his patient and a moral obligation to warn the patient's fiancée or her family. If the physician kept silent, this might make him an accomplice to the "crime."[17]

Having spent a year in Europe after completion of his medical training in New York in 1873 and having translated into English the eminent Paris venereologist Jean-Alfred Fournier's (1832–1914) treatise *Syphilis and Marriage* (1880),[18] Morrow was particularly well informed about the French literature on this issue. Although France had an absolute legal duty of medical confidentiality, opinions were divided among the relevant authors. While some were ready to breach patient confidence in this situation, others insisted on keeping the medical secret or would disclose only with the patient's consent. The Paris physician Georges Thibierge (1856–1926), for example, had made the point that keeping strict confidentiality was in the interest not only of the individual patient but of society at large. If syphilitic patients could no longer rely on their doctor's secrecy, they would seek instead the help of a charlatan or quack, resort to self-treatment, or forgo any treatment—to the detriment of society as well as the individual. Morrow respected this line of argument.[19] In the end, he agreed that medical secrecy should be kept on the basis of the utilitarian principle of "the greatest good to the greatest number" but that a physician who made the conscientious decision to breach the confidence of a stubborn and irresponsible VD patient should not be blamed.[20]

Morrow also mentioned a proposal made by the Paris professor of legal medicine Paul Brouardel (1837–1906), on the basis of one of his cases, to use a subterfuge by suggesting to the bride's father that his future son-in-law should take out a life insurance policy. The young man would refuse to do so, expecting that the required medical examination would reveal his venereal disease. The bride's family could then draw their own conclusions from his behavior. However, the health checks made by the American insurance companies were too unreliable, in Morrow's opinion, to follow this example.[21] Another French suggestion was that doctors should issue certificates confirming the absence of VD before a marriage could take place. Yet Morrow noted the "hostility of public sentiment" in the United States against such a restrictive measure, though efforts had been made in Michigan and some other states to issue such regulations. On a practical level, he thought, they would be ineffective, as questionable offices for the sale of such certificates would be opened, or people would evade restrictions by crossing the border to a state that did not have such laws.[22]

The problem that medical secrecy could lead to situations in which women unknowingly entered marriage with syphilitic men also featured in the fictional literature of the period. The American feminist writer Charlotte Perkins Gilman (1860–1935), who was a supporter of Morrow's campaign for the eradication of venereal disease, included the issue in her novel *The Crux* (1911), a story about the emancipation of a young woman, Vivian Lane. In one scene, the young heroine's female reformist physician, Dr. Jane Bellair, discusses the matter with Dr. Richard Hale, the traditionally minded doctor of Vivian's syphilis-infected fiancé, Morton Elder. Dr. Hale refuses to disclose Morton's disease, pointing out that, as a matter of professional honor, he will not betray his patient's confidence, to which Dr. Bellair sarcastically replies, "A man's honor always seems to want to kill a woman to satisfy it."[23] Breaking with (male-dominated) medical etiquette, Dr. Bellair warns Vivian, saving her from the fate of infection by Morton. Remarkably, Vivian ends up becoming engaged to Dr. Hale, an indication that his misogyny will be overcome.[24] In this way, Gilman criticized male notions of medical professional honor that could result in double standards regarding men and women when it came to venereal diseases. Beyond this, it was her general aim that women should become better informed about matters of sexuality and health, which would enable them to lead self-determined lives as well as to bear and raise healthy, strong children. This latter aspect reflected Gilman's affinity to contemporary eugenic ideas, which she shared with Morrow.[25]

Morrow further discussed the more frequent situation of a venereal infection *after* marriage, with the disease usually thought to have been introduced into the household by the husband who had visited a prostitute or had an extramarital affair. Doctors' common professional strategy was then to provide treatment for the wife without informing her of the exact nature of the disease, thus protecting the husband's confidence and avoiding domestic consequences. However, Morrow knew that this approach was, in practice, "extremely difficult" and that "the little comedy of deception and falsehood most often proves a dismal failure." He therefore referred to Thibierge's advice that the husband should openly admit his fault so that proper treatment of both partners could be ensured and pregnancy avoided.[26]

In 1907, Morrow, as president of the American Society of Sanitary and Moral Prophylaxis, invited the New York lawyer William Archer Purrington to speak to its members about the legal aspects of medical secrecy. Purrington used the opportunity to repeat his criticisms of the New York statute on medical privilege in court (see chapter 1) before he addressed the notorious issue of the VD-infected man who refuses to give up his intention to marry and insists on his doctor's secrecy. Drawing upon the analogy that poisoning of weapons or wells and sending out of infection were forbidden under the laws of war, he suggested that someone "poisoning the springs of life" with venereal disease had no claim to the privilege of confidentiality. Doctors were therefore legally as well as morally entitled to disclose if there was no other way of dealing with the situation. He also assured his listeners that reporting of venereal disease to the boards of health was not only permitted but commanded from the legal point of view.[27]

The conflicts faced by doctors in warning contacts of patients with venereal disease, especially in cases of impending marriage, had been discussed in the medical professional press for some time before Morrow and Gilman brought the issue to the attention of broader audiences. Alerted by the British *Kitson v. Playfair* libel case,[28] an 1896 editorial in the *Journal of the American Medical Association* warned that "every breach of confidence is done at the violator's peril" and that in venereal diseases such violations would "probably be most disastrous."[29] On the other hand, a further editorial in the same journal, just two years later, assured its readers that "no one would blame a physician if, for example, he should do his utmost to prevent, let us say, the marriage of an innocent female to a man rotten with syphilis or suffering from any form of actively contagious venereal disease."[30] While under some interpretations of the laws the doctor might become liable for damages, such cases were better known in France, where the courts gave more weight to the financial implications of marriage than in the United States.[31]

The state of Michigan, in 1899, was the first to make it a felony for persons afflicted with syphilis or gonorrhea to marry, threatening punishment with a fine between five hundred and one thousand dollars or imprisonment up to five years. Men had to swear to their health before they could take marriage vows. The new law also permitted testimony of the wife against the husband and removed the medical privilege in court proceedings for this cause.[32] The news from Michigan was welcomed in the pages of *JAMA*.[33] Various articles in the years that followed, in *JAMA* as well as in other medical journals, suggested that doctors' warning of contacts of irresponsible VD patients should be permissible and that marriages in such cases should be prevented.[34] In part, such proposals were made in a eugenic context.[35]

Nonetheless, in 1910, New York physician Norman Barnesby (1875–1946) still mentioned medical secrecy about venereal disease before marriage as a common "crime against posterity" in his book about failings and shortcomings in the profession, *Medical Chaos and Crime*.[36] However, attitudes in the organized medical profession toward confidentiality in cases of infectious disease, more generally, were soon influenced by the notorious New York case of asymptomatic typhoid carrier and cook Mary Mallon (1869–1938), or "Typhoid Mary" as she became popularly known. She had been compulsorily hospitalized from 1907 to 1910 and was eventually, in 1915, detained in an isolation hospital for the rest of her life as a danger to public health, being held to have been the source of outbreaks of the disease.[37] At the request of the Committee on Reporting Typhoid Fever of the American Public Health Association, the AMA considered in 1911–12 a revision of its 1903 "Principles of Medical Ethics" that would include a requirement for doctors to notify the health authorities of cases of contagious disease.[38] In fact, the revised "Principles," adopted by the AMA in 1912, incorporated such a notification as a physician's duty in epidemics: "At all times, it is the duty of the physician to notify the properly constituted public health authorities of every case of communicable disease under his care, in accordance with the laws, rules and regulations of the health authorities of the locality in which the patient is."[39] Yet, in more general terms, the revised "Principles" made decisions about disclosure a matter of the individual physician's conscience. After the traditional commitment to medical confidentiality, they stated, "There are occasions, however, when a physician must determine whether or not his duty to society requires him to take definite action to protect a healthy individual from becoming infected because the physician has knowledge, obtained through the confidences entrusted to him as a physician, of a communicable disease to which the healthy individual is about to be exposed. In such a case, the physician should act as he would desire another to act toward one of his own family under like circumstances."[40]

In line with a certain laissez faire attitude that characterized the AMA's ethics policies in those years, the "Principles" thus referred doctors to the Golden Rule ("to do as you would want to be done by").[41] Moreover, the AMA to some extent avoided taking responsibility for its own advice by adding a kind of disclaimer: "Before he determines his course, the physician should know the civil law of his commonwealth concerning privileged communications."[42] Thus fears that doctors might make themselves vulnerable to claims for damages, on the grounds of slander or libel, continued to be valid and would have been especially relevant in conditions as socially stigmatized as venereal diseases.

By 1914 seven American states, and by 1922 twenty states, had enacted statutes to prevent the introduction of venereal disease into marriage. However, many states classed infection of the spouse only as a misdemeanor; the laws were rarely enforced; and if they included a premarital medical examination, this pertained only to men. It was regarded as a moral insult for the bride to undergo such a medical investigation. None of the states required a serological (Wassermann) test.[43] Substantial change came only in 1935, when Connecticut was the first state to require the Wassermann test and a physical examination for all brides and grooms before a marriage certificate could be issued. Yet Morrow's prediction, some thirty years earlier, that people would evade such restrictions became a reality as well. In the years before New York passed an analogous law in 1938, the number of weekend marriages in New York counties bordering Connecticut rose significantly.[44] Still, the requirement, at least in some places, of medical certificates before marriage took some of the edge out of the conflicts that American doctors faced over confidentiality and venereal disease.

A Landmark Decision in Germany, Its Context, and Its Consequences

In March 1905, and again in June, the *Journal of the American Medical Association* reported about a German case in which a doctor had been taken to court for breach of medical secrecy.[45] As was soon to become clear, the decision in this case had a significant impact on the subsequent legal understanding of confidentiality.

A Berlin medical practitioner, Dr. L., had warned a married woman, Mrs. J., when bringing one of her children for vaccination, that her sister-in-law, who was his patient, suffered from syphilis. The doctor had learned from Mrs. J. that the sister-in-law lived in the same house and was close to the children, sometimes taking them into bed with her. He therefore advised Mrs. J. that her children, especially the newly vaccinated one, must not touch their aunt. Mrs. J. subsequently told another tenant about this conversation, and the latter distributed the news of the sister-in-law's syphilis all over the house. A few days later, the sister-in-law, accompanied by her mother, went to see Dr. L. The mother confronted him, asking what rumors he was spreading about her daughter. The doctor frankly declared that her daughter had syphilis, and heated arguments followed.[46]

The court of first instance, the Berlin District Court I, found Dr. L. guilty of breach of secrecy on two counts (disclosure to Mrs. J. and to the sister-in-law's mother) under section 300 of the Reich Penal Code. Taking into account his motive to warn of an infection, it sentenced him to the rather modest fine of twenty reichsmarks. However, the doctor appealed against the verdict, having been encouraged to do so by a prominent Berlin colleague, the psychiatrist Albert Moll (1862–1939), who had just published a comprehensive book on medical ethics.[47] Moll was aware of the German Supreme Court's (*Reichsgericht*) decision of 1903 in the protracted Hamburg divorce case (see chapter 1), which had said that a breach of patient confidentiality by warning potential contact persons might be justified on the grounds of "higher moral duties."[48] He had also corresponded on the subject with a judge, Landgerichtsdirektor Fromme in Magdeburg. Already before the Supreme Court's decision of 1903, Fromme had constructed on the basis of earlier court decisions a doctor's moral duty to disclose in the interest of the "well-being of a human being" (*Wohl eines Menschen*), including warning of infection with venereal disease.[49] In fact, as it turned out, in the Berlin case, the Supreme Court followed a similar line of argument. Dr. L.'s professional duty of "conscientious practice" according to the Prussian Law on Medical Courts of Honor (1899), the Supreme Court argued, implied a duty to warn patients in danger of being infected by people in their close environment. The doctor might even have become liable for negligent physical injury under section 230 of the Reich Penal Code if he had not warned the children's mother. His first disclosure was therefore justified. Also, the second disclosure, to the sister-in-law's mother, was not punishable in the Supreme Court's opinion: the "secret" was by then widely known, and since the two women had come together to see Dr. L., he could assume that he could speak freely with his patient's consent.[50] The verdict of the court of first instance was lifted, and the case was referred back to the lower court. The Berlin District Court II subsequently acquitted the doctor.[51]

As in the American case of *Simonsen v. Swenson* some fifteen years later, concern for the health of third parties had thus trumped the duty of confidentiality. Looking back on the Berlin trial, Moll commented in 1936 that it had led to a real change in German jurisdiction in this area. Previous decisions had supported a tendency that would have led to a situation where doctors, bound to the legal duty of secrecy, had to look on while their patients infected others with venereal disease.[52] The issue was known beyond medical circles. As in America, the moral problem for a doctor who knew about a VD patient's marriage plans but felt unable to betray confidentiality and to warn the fiancée had also become a subject of literary fiction. In her 1902 novella *Schweigen* (Silence), Ottilie Franzos (1856–1904), writing under her pen name, F. Ottmer, told the story of a doctor who refrains from informing a young woman about her future husband's infection, believing that his professional honor obliges him to preserve medical secrecy. His decision has dire consequences: following marriage, the woman gets infected, and the couple's baby dies shortly after the husband has succumbed to his illness.[53]

In a real case in 1903, a syphilis-infected wife sued her father-in-law for damages, as he had consented to the marriage despite knowing about the disease of his son (who had meanwhile died). The father-in-law justified himself by declaring that his son's doctor had assured him that his son had been cured and was able to marry. The wife maintained that the opposite was true. The doctor, referring to his legal duty of secrecy and his entitlement to refuse giving evidence in court (see chapter 1), was unwilling to testify—a decision that was endorsed by the Supreme Court because only the husband would have been able to waive confidentiality.[54] The problem of this case was thus the doctor's legally justified insistence on medical secrecy, which prevented clarification of the situation. The liability in civil law of a husband who had knowingly infected his wife with a venereal disease was uncontroversial.[55]

In 1904, Siegfried Placzek (1866–1946), a Berlin forensic psychiatrist and expert on medical confidentiality, still advised his colleagues, if they wanted to avoid getting in conflict with the law, to keep strict silence in situations where a future spouse had VD.[56] The Leipzig physician Wilhelm Rudeck, writing on medico-legal aspects of marriage, shared this view but recommended Brouardel's trick of demanding a life insurance policy for the future husband.[57] The common legal opinion at this time was that a doctor was not permitted to disclose the venereal disease of a groom to the bride. The honorable motive to warn did not count.[58] The Supreme Court's 1905 verdict in the Berlin case opened a way out of this dilemma by offering the professional duty to warn contacts of an infected patient as a justification for breaching medical confidentiality. Its legal and medical significance was immediately recognized,[59] although the Supreme Court's line of argument was criticized by some commentators. It was questioned, for example, whether the professional duty of conscientious practice stated in a *Prussian* law (that is, the Law on Medical Courts of Honor) could override the duty of professional secrecy required by a *Reich* law (that is, section 300 of the Penal Code).[60] Moreover, it was argued that a duty of conscientious practice could never entitle a physician to undertake a punishable act.[61] It was also pointed out that the correct way of dealing with the situation would have been for Dr. L. to first discuss with his patient the danger of infecting others and to ask her to release him from his duty of confidentiality before warning contact persons. If such permission were refused by the patient, he would still have been morally justified to warn, though he would have had to accept a small fine for breach of secrecy.[62]

However, the Supreme Court's landmark decision of 1905 stood in a much broader context of debates on the notification of infectious diseases more generally. In principle, the Prussian Sanitary Regulations of 1835 already required medical practitioners, as well as landlords and heads of households, to report cases of dangerous contagious diseases to the police. However, syphilitic infections only had to be reported (including the patient's name) when keeping the disease secret would have, in the doctor's opinion, negative consequences for the patient concerned or the community and, under the Prussian General Law (1794), in cases of prostitutes working in brothels. Moreover, if civil doctors treated soldiers for syphilis, they had to report them to the relevant military commander or senior medical officer. In addition, all medical practitioners and heads of hospitals were obliged to provide anonymously, in their quarterly sanitary reports, the number of patients with syphilis they had seen or treated.[63] However, many doctors ignored these regulations, fearing a loss of clientele if they disclosed patients' venereal diseases. By the late nineteenth century, these old regulations had been virtually "forgotten," so a ministerial decree in 1898 had to remind doctors that they were actually still valid.[64]

In the aftermath of a devastating cholera epidemic in Hamburg (in 1892), the German Reichstag passed in 1900 the Law on the Combating of Diseases Constituting a Public Danger (*Gesetz, betreffend die Bekämpfung gemeingefährlicher Krankheiten*), which obliged doctors to immediately report to the police any case of suspected illness or death from cholera, typhus, smallpox, leprosy, yellow fever, or plague.[65] The implementation laws of the German states could extend this list. For example, in 1905, Prussia added, among others, diphtheria, meningitis, scarlet fever, and typhoid fever as notifiable diseases and required the reporting of deaths from pulmonary and laryngeal tuberculosis.[66] Venereal diseases, however, were left out of the list of generally notifiable diseases. Only prostitutes infected with VD could, under police ordinances, be subjected to surveillance, isolation in a hospital, and compulsory treatment.[67] A requirement for civil doctors to report VD-infected soldiers to the military authorities was rejected by the lower house (*Abgeordnetenhaus*) of the Prussian Parliament after substantial debate. It was thought that such a requirement would result in soldiers failing to seek proper medical treatment under these circumstances, instead resorting to treatment from quacks, or civil doctors refusing to treat soldiers.[68]

In this context, lively public discussions took place about whether venereal diseases should be made *generally* notifiable, like other dangerous infections. Many doctors were opposed to such measures, arguing that disclosure in such socially sensitive diseases would destroy their patients' trust, lead to concealing of infections, or make patients resort to unqualified lay healers. Some critics thought that the existing laws already provided sufficient powers to deal with VD patients who negligently or deliberately spread their infections.[69] Similarly to Morrow in the United States, some prominent German venereologists, including Albert Neisser (1855–1916) of Breslau (the discoverer of the gonorrhea bacterium) and Alfred Blaschko (1858–1922) of Berlin, founded in 1902 the German Society for Combating Venereal Diseases (*Deutsche Gesellschaft zur Bekämpfung der Geschlechtskrankheiten*), which united social reformers with various professional and political backgrounds.[70] In March 1905, just two months before the Supreme Court issued its decision in the Berlin case, the German Society, which by then had about four thousand members, held its second congress in Munich, with "Medical Secrecy and Venereal Diseases" as one of its main themes. As the *Journal of the American Medical Association* reported, the "congress had a crowded attendance. [. . .] [A]udiences included the general public, with a very large official and legal attendance."[71] The papers read at this occasion reflected the wide spectrum of opinions even among anti-VD campaigners. The most radical proposals were made by the Frankfurt police surgeon Max Flesch (1852–1942 or 1944), who advocated a general duty for doctors to report all VD cases to the authorities, compulsory hospitalization of renitent patients, and reporting of some cases to the courts. In order to prevent VD patients from switching to treatment by quacks, he suggested that lay healers should likewise be required to notify the authorities.[72]

Neisser, being away for syphilis research on apes in Java, had his paper read out at the congress by a Breslau colleague, Martin Chotzen. His more moderate proposal argued for a right (as opposed to a duty) of doctors to report "careless and frivolous" VD patients to the health authorities. Such medical reporting, rather than notification to the police, would be more acceptable to doctors. He also suggested amending section 300 of the Penal Code by entitling doctors to disclose private information in order to prevent infection and harm to other people. In the same vein, he proposed that judges should be enabled to force doctors to disclose confidential details on their patients if this was essential for the court's decision making—a suggestion that reflected the British situation (see chapter 1). In Neisser's view, public interests had to prevail over individual interests "under all circumstances."[73]

Max Bernstein (1854–1925), a prominent Munich lawyer, disagreed, however, with Neisser's proposals, arguing in his paper that there were already sufficient exceptions from the legal duty of medical secrecy—for example, in crimes (section 139 of the Penal Code), in the infections falling under the law of 1900 on dangerous contagious diseases, and in all cases in which the patient had consented to disclosure. Importantly, he pointed to the Supreme Court's decision in the Hamburg divorce case of 1903, permitting doctors, in principle, on the grounds of a higher moral duty, to breach medical confidentiality to protect potential contact persons from infection with venereal disease. Bernstein also wished to maintain the status quo for German doctors when giving evidence in court. It was to be left to their discretion in each case whether they wanted to testify on private patient details or to remain silent. If the patient had not waived confidentiality and the doctor refused to give evidence, the judge could draw his own conclusions from this.[74]

Given the divergence of opinions that emerged at this Munich congress, it is unsurprising that the Supreme Court decision in the Berlin case also attracted political attention. In March 1906, it became a subject of debate in the upper house (*Herrenhaus*) of the Prussian Parliament, when Count von Hutten-Czapski stated that it was "apt to undermine patients' trust in their doctor and make them turn to quacks." He asked the Minister for Educational and Medical Affairs to ensure that any future cases of breach of professional secrecy should be subjected to disciplinary proceedings at the medical courts of honor.[75] The director of the ministry's medical department, Adolph Förster (1847–1919), replied that he largely shared the count's concerns. The doctor–patient relationship, which was a contractual relationship based on trust and confidentiality, had been "seriously shaken" by the Supreme Court decision so that some patients might now hesitate to consult a doctor in confidential matters. The decision gave doctors a right to disclose, at their discretion, information from their practice in order to avert harm from other persons. Förster acknowledged the public health argument in favor of such disclosure but hoped that the Supreme Court might correct its position in this matter and that future penal law reform might find a better balance between the needs of the public and the relationship of trust between doctor and patient.[76] However, the Supreme Court did not change its opinion, and Förster himself endorsed the new line in 1907 as chairman of the central appeal court of Prussia's medical courts of honor when he acquitted a doctor who had reported a syphilitic teacher to the Inspector of Schools. The court of honor accepted that this doctor had experienced a collision of his duty of medical secrecy and his duty to warn of a danger to public health (that is, in this case, the health of schoolchildren) and that his decision to report had not violated the reputation and honor of the medical profession.[77]

The German Supreme Court decisions of 1903 and 1905, and the 1907 decision of the Prussian Medical Court of Honor, thus established a line of thought on confidentiality that placed the individual's interest in medical secrecy below the social interest in preventing harm to third parties and in public health more generally. German doctors' attitudes toward disclosure were slow to change, however. When, in the context of increasing VD infections during World War I, the social insurance organizations (*Landesversicherungsanstalten*) established advice centers for VD patients and doctors were supposed to report such patients by name and via their health insurance to these centers, this was met with concerns from the medical profession. As an editorial in the journal of the German Medical Association, the *Ärztliches Vereinsblatt für Deutschland*, complained in 1916, "Nowadays one uses every opportunity to put forward the higher moral duty of public health, to which the duty of secrecy is supposed to subordinate itself."[78] In December 1918, the Reich Office for Economic Demobilization issued an order to report soldiers with VD to the social insurance organizations.[79] In the Weimar Republic, the advice centers became instruments of VD control as well as social welfare. The majority of patients attending these centers had been reported to them by a doctor, hospital, health insurance organization, or, in the case of soldiers, by the military administration; only 30 to 40 percent of patients visited them voluntarily.[80] Yet the medical disciplinary tribunals remained watchful. In 1925, the central Prussian Medical Court of Honor in Berlin confirmed the disciplinary warning of a doctor who had reported two women as sources of venereal disease to the public health bureau of a health insurance organization without having direct evidence of their infection, just relying on information from a patient.[81]

Eventually, in 1927, a Reich Law for Combating Venereal Diseases (*Reichsgesetz zur Bekämpfung der Geschlechtskrankheiten*) required that all VD-infected persons had to be treated by qualified medical practitioners (as opposed to lay healers). Doctors were now legally obliged to report to the health authorities or advice centers any VD patients who did not comply with the treatment plan or failed to submit to medical controls or who posed a danger to others through their personal circumstances or occupation. If necessary, the health authorities could then order compulsory hospitalization of the patients concerned. The same law bound the staff of the health authorities and of the VD advice centers to confidentiality, though they were entitled to disclose a patient's venereal disease with the relevant medical officer's consent to other officials or other persons who had a legitimate health interest in being informed.[82] As a commentator from the Reich Justice Ministry explained in 1928, the basic idea of this law had developed from the Supreme Court decision of 1905 in the Berlin case, which had become the "paradigmatic case for the whole doctrine" that the individual's interest in medical secrecy had to stand aside if prevailing interests of the community required this.[83] Within the first year of the VD law's operation, the Berlin health authorities alone were notified about two thousand cases in which patients had not completed their cure.[84]

As in America, premarital health examinations were also considered. From 1921, couples planning to marry had to obtain a leaflet from the registrar's office advising that those who were infected with a venereal disease should not enter marriage, and the law of 1927 made it compulsory for each partner to inform the future spouse of an earlier infection. Obligatory medical certificates before marriage were introduced in Germany in 1935 by the National Socialist regime as part of its wider agenda in eugenics and racial hygiene. However, due to personnel shortages and organizational problems, this certificate system was only partially implemented.[85]

Comparing Germany to the United States

Germany and the United States thus followed similar trajectories in the issue of medical confidentiality and venereal disease. The traditional view that doctors should observe strict secrecy regarding patients with VD was increasingly criticized and gradually gave way to disclosure and reporting where this was required by public health interests. Pressure came from social reformers as well as state authorities. Landmark decisions of the higher courts that condoned disclosure in cases of venereal diseases prepared the ground for regulations to this effect. The general strategy to fight the spread of dangerous infectious diseases by reporting and surveillance eventually prevailed in the case of VD over the resistance of those in the medical profession who insisted on the inviolability of medical secrecy as the bedrock of the doctor–patient relationship. In short, public health increasingly came before the privacy interests of individual patients. The health authorities in Germany and the boards of health in the United States carried out disease surveillance and were entitled to intervene if public health was perceived to be at risk. They were the tangible, institutional expressions of a policy of state control of VD that had become established in the aftermath of the First World War.

British Discussions on Medical Secrecy and Venereal Diseases

In Britain, as in late nineteenth- and early twentieth-century America and Germany, questions arose about whether physicians had a general duty to report cases of venereal disease to the health authorities. As historians of these debates such as Roger Davidson,

Lesley Hall, and Lutz Sauerteig have emphasized, England took a distinctly "voluntarist" route that counted on free access to confidential diagnosis and therapy instead of notification and compulsory treatment as the strategy for fighting venereal disease. Besides the general influence of liberal traditions in England, the specific history of the English Contagious Diseases Acts of 1864, 1866, and 1869 appears to have been a main factor in this development. Due to strong protests, especially from the early women's movement and the churches, these acts, which had provided for compulsory medical inspection of prostitutes in certain garrison and naval towns, were suspended in 1883 and repealed in 1886. In their critics' view, they had constituted an unacceptable intervention of the state into matters of health and had enshrined in law a double standard that targeted female prostitutes but left their male clients untouched.[86]

In the 1880s and 1890s, attempts to introduce compulsory notification of venereal diseases through legal powers of local authorities and provisions of the Infectious Diseases (Notification) Act of 1889 failed, not least because of opposition from the medical profession. Doctors argued that reporting of venereal diseases would damage the confidential and fiduciary physician–patient relationship that was necessary for effective treatment. Notification would drive VD patients away from qualified medical practitioners and would therefore be counterproductive. It was also seen as an infringement of the individual patient's right to privacy. Behind much of this argumentation was private doctors' concern for maintaining professional status and financial independence.[87]

A closer look at discussions of the time reveals, however, divided opinions within the medical profession on other aspects of the issue of venereal disease and confidentiality, particularly regarding the warning of contacts. For example, Robert Saundby (1849–1918), vice president of the Council of the British Medical Association and a member of the General Medical Council, in his book *Medical Ethics* (1907), advised colleagues that while they should strongly object to the marriage plans of a patient with an infectious disease that might be transmitted to the future spouse and children and urge him or her to take all necessary precautions for preventing infection, the disease should not be disclosed to the endangered persons without the patient's consent.[88] In contrast, Campbell Williams, a fellow of the Royal College of Surgeons of England, in a lecture published in the *Lancet* in 1906, not only suggested the reintroduction of inspection and compulsory hospital treatment of prostitutes on a larger scale but also raised the question of whether a breach of professional secrecy was justified in order to prevent an innocent person from being infected.[89] While his proposal to, in effect, reinstate the Contagious Diseases Acts, now for the whole country, led to a lively correspondence with readers who objected to the targeting of female prostitutes to protect men's sexual desires, the suggestion to warn contacts was not contradicted.[90] In Williams's view, it was morally legitimate, for example, to warn the wife of a VD-infected husband (even if this led to divorce) or the employer of a servant or nursemaid who had contracted syphilis or gonorrhea.[91] A similar position on the latter point was taken by another fellow of the Royal College of Surgeons, Hunterian Professor J. Howell Evans. In an address to the Medico-Legal Society in London in 1908, he suggested that if the employer had paid the medical attendant, he was entitled to receive correct information on the servant's illness.[92]

In practice, however, such cases were controversial, and some patients sued their doctors for slander or libel. The secretary of the Medical Defence Union, A. G. Bateman, recounted, for example, the case of a London fireman who had (unsuccessfully) sued the medical officer of the fire brigade for libel because he had reported his infection with venereal disease to the employer, resulting in the man's dismissal.[93] In another case, a Leeds doctor who, after examination of a barmaid, had communicated his diagnosis to her employer and the housekeeper, as well as to her (alleged) husband and a fellow barmaid, was taken to court by the patient. While acquitted regarding the first three communications, which were held to be privileged, he was convicted to pay £75 in damages for libel because of the fourth communication.[94]

Concerns about an increase of venereal infections during the First World War led to official steps against the spread of the disease.

The report of the Royal Commission on Venereal Diseases in 1916 recommended voluntary treatment over notification and compulsory measures. On this basis, the Local Government Board under the Public Health (Venereal Disease) Regulations of the same year introduced a system of VD treatment centers that guaranteed confidential and free diagnosis and therapy. By the end of 1920, the local authorities in England and Wales had established 185 such centers.[95] The 1916 Public Health Regulations also formed the basis of VD policy in Scotland. Although Scottish health authorities and civic leaders campaigned in the 1920s for legal powers to notify, detain, and penalize VD patients who did not comply with their prescribed course of treatment, such proposals were resisted and frustrated by the Ministry of Health in London.[96] Venereal diseases were not made notifiable, and voluntary treatment (supported by health education) remained the main approach in Scotland as well as in England. The only VD-related condition that had to be reported, in England and Wales, was ophthalmia neonatorum—that is, gonorrheal eye infections in newborn children.[97] Education of the public on VD was especially the task of the National Council for Combating Venereal Diseases, a semiofficial organization founded in 1914 that was financially supported by the Local Government Board and later the Ministry of Health and local authorities.[98] The National Council also contemplated advocating a policy of health certificates for both parties before marriage, and it advised parents to inquire about the health of their son's or daughter's future marriage partner.[99]

Confidentiality was a key element of the voluntary treatment scheme, guaranteed in Article II, Section 2 of the Public Health (Venereal Disease) Regulations of 1916.[100] Unsurprisingly, therefore, the much-publicized divorce trials of *Garner v. Garner* (1920) and *Needham v. Needham* (1921), in which doctors from VD treatment centers were forced by the judge to give evidence in court about their patients' illness (see chapter 1), caused considerable concern. For example, referring to the case of *Garner v. Garner*, the *Times* noted, "It is generally accepted that Mr. Justice McCardie was legally correct in his ruling, but the opinion is expressed that the decision will lead to a setback in the campaign to lessen the serious prevalence of foul disease."[101]

The *Lancet* feared that the public's attendance of the treatment centers would diminish if the promised confidentiality was not strictly kept and believed that McCardie's decision to compel the doctor to testify demanded a change in the law.[102] Similarly, the *Daily Chronicle* commented with regard to the Needham case, in which Justice Horridge had ordered Dr. John Elliott of the Chester Venereal Clinic to give evidence about the disease of his patient, "It is clear that if there is no guarantee of professional secrecy in certain kinds of clinic the whole object of the Ministry of Health acting in the interests of the public is likely to be defeated. The matter requires legislation."[103]

The Ministry of Health, worried about the impact of these legal cases on the voluntary VD treatment scheme, tried to lobby the Ministry of Justice and the Lord Chancellor for legislation that would grant a medical privilege in court for civil proceedings. However, especially the Lord Chancellor, Viscount Birkenhead, successfully resisted this proposal with the argument that such a privilege for doctors would obstruct the administration of justice, and the status quo in this matter continued unchanged.[104]

Medico-legal opinion in the interwar period remained divided on the question of whether contacts of a VD patient should be warned even without his or her consent. W. G. Aitchison Robertson, lecturer in medical jurisprudence and public health in Edinburgh, suggested in his book *Medical Conduct and Practice* (1921) that the doctor should inform the father-in-law in the case of a syphilitic fiancé who refuses to disclose his illness. The breach of the ethical rule of secrecy would occur in the interest of others, and it was inconceivable that a court of law would hold the doctor culpable in such a case.[105] On the other hand, Hugh Woods, secretary of the London and Counties Medical Protection Society, advised in 1924 in the *Lancet* that in this situation it was not the doctor's duty to provide information about the patient's condition. If it were "right and proper" to do so, this ought to be "openly authorised by the rules of the medical profession." Polemically, he asked, if it was justifiable to divulge the fiancé's syphilis, then why not disclose if he was a drunkard or otherwise objectionable? Acting in this way would soon end the medical practitioner's career and do "vastly more harm than good."[106]

By contrast, speaking from a legal perspective, Lord Riddell (George Allardice Riddell, 1865–1934) suggested in a lecture for the Medico-Legal Society in 1927 that a doctor might be able to claim a right to disclosure on the grounds of a "public duty to prevent wrongful acts." There would be occasions when the duty of professional secrecy should be "subordinated to the higher duties of common humanity." As he further explained, "The terrible consequences to innocent persons of secrecy in the hypothetical cases of the syphilitic husband or wife, the diseased fiancé, the syphilitic cook or nurse [. . .] are matters deeply concerning the public welfare. Disclosure to avoid such consequences is justifiable and perhaps obligatory on both legal and ethical grounds."[107]

Riddell stopped short, however, of providing doctors with a definite rule of conduct: all depended on the conscience of the individual medical practitioner and the particular circumstances of the case. As he noted, there was no English legal case that could serve as a direct precedent. However, Riddell did mention a 1921 Hungarian case in which a VD specialist had prevented one of his patients, whom he had recently treated for fresh syphilis, from entering a public bath by reporting him to the attendant. The patient had subsequently sued the doctor for breach of professional secrecy, but the court had dismissed the case, as the doctor had "acted in the interests of the community."[108]

The London consultant F. G. Crookshank (1873–1933), a fellow of the Royal College of Physicians, took a similarly relativist view in a handbook chapter on the medico-legal aspects of venereal diseases in 1931. If confronted with the situation of the syphilitic fiancé, the doctor should, after explaining the position to the patient, follow his personal judgment and conscience and be prepared to face the consequences of whatever line of action he decided to pursue.

He also suggested that the doctor should consult his professional defense society (that is, the London and Counties Medical Protection Society or the Medical Defence Union). Crookshank's personal opinion tended, however, toward disclosure: the health and happiness of innocent persons should not be jeopardized by the doctor's own timidity.[109] This stance matched contemporary legal advice. As London barrister D. Harcourt Kitchin put it in his handbook *Law for the Medical Practitioner* (1941), "If [. . .] the disclosure was to protect the health of others—*e.g.*, wife, fiancée or fellow-workmen—the court might take the view that a moral duty to disclose constituted a good defence."[110]

Another strategy to prevent the spread of venereal diseases was to trace the past sexual contacts of VD patients to identify and treat persons who were sources of infection. In Scotland, such contact tracing was informally adopted by VD treatment centers in the 1930s through community nurses, who also followed up on patients who had defaulted from treatment. Patients were asked to give each contact person a letter containing, in medical jargon, relevant details of the patient's condition (without actually mentioning the term "VD") and to tell him or her to present the letter to a doctor at one of the treatment centers.[111] Such schemes, however, carried the risk of legal action for slander or for "injury done to feelings" if the contact person turned out to be free of infection, and they operated rather discreetly in the knowledge that they conflicted with medical confidentiality.[112]

Conclusions

In all three countries—Britain, the United States, and Germany—official notification of venereal diseases and the warning of contacts of VD patients, without patients' consent or even against their wishes, were major topics of medical and legal debate. The approaches that were taken in these issues differed to some extent among these countries. While by 1919 all American states had introduced obligatory reporting of venereal cases to the health authorities, at least anonymously or with the patient's initials, Britain went a different route with the Public Health Regulations of 1916, which offered free access to voluntary, confidential treatment in VD clinics. Only venereal infections in newborn children were made notifiable. In 1927, after long discussions, Germany introduced selective notification only in the case of VD patients who defaulted from treatment or whose occupation (e.g., as prostitutes) posed special risks. Thus a model of state intervention prevailed in Germany as well as in the United States, whereas a liberal approach to the VD problem characterized the British "voluntarist" strategy. In all three countries, the medical profession's opposition to notification followed a similar line of argument: reporting the patient's venereal disease would destroy the confidential nature of the doctor–patient relationship, which was regarded as crucial for effective treatment; patients would be driven away from duly qualified medical practitioners into the hands of unscrupulous lay healers, or they would resort to self-medication or forgo any treatment, which would result in social as well as individual damage.

Similar arguments were made against the warning of contact persons without the patient's consent. The issue was widely discussed not only in medical and legal professional circles but in broader society. In Germany, a landmark decision of the Supreme Court in 1905 set the precedent for disclosure in order to avert harm from the VD patient's family or other persons in his or her social environment. In the United States, the Supreme Court of Nebraska set a similar precedent in 1920, through the case of *Simonsen v. Swenson*. Britain lacked such a precedent, which led to considerable uncertainty for doctors, as they might have to face actions for libel or slander if they disclosed their patients' venereal disease to others. Medico-legal advice on the matter was therefore contradictory or noncommittal, although moral arguments predominantly advocated disclosure in order to prevent infection of "the innocent." In the specific scenario of the syphilitic fiancé, a breach of confidentiality was widely seen as justifiable, though the risk of disclosure remained with the doctor concerned. Premarital health certification as a potential solution to the problem was considered in all three countries but was only (to some extent) implemented in the United States and Germany, in part driven by eugenic aims.

With its common-law rule of rejecting a medical privilege in court, Britain encountered special difficulties with the officially guaranteed confidentiality of its voluntary VD treatment scheme when doctors were compelled in divorce trials to give evidence regarding the illness of one of the spouses. While in the early 1920s the issue caused disquiet in the British medical and daily press, as well as in the Ministry of Health, the judiciary successfully resisted any change in the question of a privilege in court: doctors had none, and still have none to the present day. In all three countries, the principle prevailed that the public interest, be it in justice or in health, had to come before private interests in confidentiality. The next chapter will further explore this theme of disclosure in the public interest for a particularly sensitive problem in the nineteenth and early twentieth centuries: illegal abortions.

Chapter 3

Abortion: Reporting a Crime or Preserving Confidentiality?

Introduction

In March 1887, an editorial in the *Journal of the American Medical Association* quoted the following suggestion of an anonymous correspondent: "When we shall have passed a law binding all physicians, upon their honor, to make public the desire of any person who may ask the performance of an abortion, we can then hope to check, in a measure, the wholesale murder of the newborn. Secrecy in regard to patients who come to us for advice should be sacredly observed, but the audacious insult of requesting an honorable physician to procure an abortion should be publicly resented by openly reporting the party."[1] The editor disagreed with this proposal: only a small proportion of women wishing to terminate a pregnancy consulted a physician, and when warned of the "physical and moral" dangers of the intervention, many of them would abandon the plan; thus it would be "unjust" to report such cases, and physicians should not be compelled to do so. Moreover, such reporting would hardly deter those who persisted in seeking the services of an abortionist. Finally, usually no reliable witnesses were present when abortion requests were made, and if doctors reported such incidents, they risked being sued for libel.[2]

This brief juxtaposition of opposing views reflects some of the problems the medical profession encountered in relation to abortion and medical confidentiality. These problems were not confined to the scenario in which a woman asked a physician for a termination of pregnancy. The question of whether to report or keep silent also arose for doctors called to attend to women who developed complications after illegal abortions, especially when those women were dying as a result of those abortions. Furthermore, the issue of medical confidentiality became relevant when doctors were asked to give evidence as witnesses in court cases on abortion. The matter was not confined to the United States, either: it was also discussed in Britain and Germany, where abortion had likewise been criminalized. In this chapter, I will analyze the discussions about confidentiality in abortion cases as they developed in the late nineteenth and early twentieth centuries in these three countries against the background of more general debates on the morality, legality, and social implications of abortion.

"The Great Crime" in Nineteenth-Century America

From the late 1850s, physicians linked to the American Medical Association had campaigned against abortion. Led by the Boston obstetrician and gynecologist Horatio Robinson Storer (1830–1922), they sought changes to the states' criminal codes that would outlaw abortion at any stage of gestation, from conception onward.[3] Neither the law nor public opinion then regarded termination of a pregnancy before the time of "quickening" (that is, the occurrence of the first fetal movements as perceived by the mother in the fourth or fifth month of pregnancy) as a criminal offence. In popular perception, only with quickening did the fetus become "animated" and truly "alive." Legally, if the woman had not been injured by the abortion procedure, whether by use of instruments or abortifacient drugs, and if the intervention had taken place before quickening, there were no grounds for an indictment.[4] Some doctors disagreed with this view, pointing out that the fetal heartbeat could be heard with the stethoscope before the time of quickening and referring to evidence on early embryonic development that had been revealed by microscopic studies undertaken by anatomists. For these doctors, early, prequickening abortion was as immoral as a late abortion or infanticide.[5]

Moreover, there were professional and political reasons for the medical antiabortion campaign. Abortion had become a lucrative business for a variety of irregular health care practitioners, midwives, and sellers of abortifacients, as abortion rates had increased in the United States since the 1830s. By mid-century, it was estimated that one of every five pregnancies was terminated. In trying to criminalize prequickening abortion, regular doctors sought state sanctions against their irregular competitors as well against those within their own profession who performed the procedure without the only then generally accepted medical reason: to save the pregnant woman's life.[6] The regular physicians' crusade for stricter abortion laws opened up the possibility for them to regain authority in questions of social policy—something they had lost with the repeal of the medical licensing laws earlier in the century.[7] Doctors' antiabortion stance was further motivated by nativist sentiments. The practice of abortion was especially prevalent among married, white, native-born, Protestant women—a fact that raised fears of society becoming overwhelmed by large families of Catholic immigrants. The massive loss of lives during the Civil War and a perceived need to repopulate the country provided an additional argument for doctors to fight abortion. Many of them even feared that "their own women" might abandon their traditional roles in the domestic sphere and as caring mothers if the practice of abortion went on unchecked.[8]

The regular physicians' campaign also gained support from the camp of their professional rivals, the homeopaths. Edwin Moses Hale (1829–1899), professor of materia medica and therapeutics at Hahnemann Medical College in Chicago, published a pamphlet in 1867 titled *The Great Crime of the Nineteenth Century*, which strongly attacked abortion on moral as well as medical grounds and made proposals for suppressing the practice through tough legislation.[9] In Hale's opinion, not only the abortionist but also the pregnant woman should be punishable, as she was "often more guilty than the person inducing the abortion, for she may, by various improper means, as bribes, threats and other inducements, influence the physician or other person to commit the crime, when his better judgment and principle would revolt against it."[10] Moreover, Hale thought the circle of aiders and abettors of the "crime" of abortion should be held responsible, which could include the husband or other male partner, the manufacturers and sellers of abortifacient drugs and instruments used for abortion, and the publishers of newspapers and magazines carrying advertisements for abortion-causing nostrums. In Hale's view, even those who merely gave advice on abortion procedures without becoming active themselves—for example, female friends, nurses, or physicians—"should be brought to justice."[11]

Hale further discussed the crucial question of whether persons who knew about an abortion but did not report it to the "proper authorities" should be punished as well. This question was especially relevant to physicians and nurses. Hale thought they should be fined and lose their licenses for a period of time if they failed to report an abortion, having in this way become "accessories after the fact."[12]

Such ideas were, as Hale himself admitted, to some extent "utopian,"[13] but the medical campaign against abortion actually did achieve a considerable tightening of the state laws during the last third of the nineteenth century. As historian James Mohr has shown, between 1860 and 1880, at least forty antiabortion statutes were introduced in American state and territorial legislation. In most states, the legal distinction between prequickening and postquickening termination of a pregnancy was abandoned, and abortion was made a crime regardless of the stage of gestation at which it was performed. In 1860, Connecticut had set the tone for this legal development by making abortion, without reference to quickening, a felony punishable by a fine of up to $1,000 and imprisonment up to five years; by also including as punishable those women who had made an abortion attempt on themselves; and by setting fines between $300 and $500 for the dissemination of information on abortifacients and abortifacient materials.[14] By 1900, virtually every jurisdiction in the United States had criminalized most types of abortion.[15]

However, public attitudes toward abortion had not really changed. Many people continued to regard early abortion (i.e., before quickening) as harmless and morally justifiable, many doctors were sympathetic to abortion requests, and terminations of pregnancy were frequently provided by regular physicians.[16] The authorities tended to invoke abortion laws and take action only if a woman had been seriously harmed or killed through the intervention, and juries were reluctant to convict abortionists.[17] Following a period of leveling off in abortion rates toward the end of the nineteenth century (probably due to a wider use of contraceptive techniques rather than fear of the law), abortion rates gradually increased from 1900 onward, with peaks shortly after World War I and then during the Great Depression. It has been estimated that by the mid-1930s, there was one abortion for every four pregnancies in the United States—that is, abortion occurred at a similar rate as in the middle of the nineteenth century.[18] Some physicians started a second antiabortion campaign in the early twentieth century, this time focusing not on legislative changes (which had largely been achieved) but on supporting law enforcement against abortionists, controlling the work of midwives, and bringing about a change of attitude among women, the general public, and the medical profession.[19] It is against this historical background and trajectory that American discussions of medical confidentiality in relation to abortion need to be seen.

Evidence, Reporting, and Confidentiality in American Abortion Cases

The existence in many states of a medical privilege in court (see chapter 1) and the professional ethical duty of confidentiality caused considerable difficulties when medical evidence was sought in legal proceedings for criminal abortion. Addressing the section on medical jurisprudence at the 1888 Annual Meeting of the American Medical Association, Iowa physician H. C. Markham argued that as long as there were no changes in the professional rules regarding the giving of evidence in abortion cases, little success could be expected in their prosecution. Therefore, in contrast to other situations where the physician was obliged to defend his patient's secret, in cases of abortion, disclosure was justifiable. As Markham put it, "Foeticide no more entitles the patient to this secrecy and confidence than does small-pox or other danger to the public, the stamping out of which is the duty of medicine to perform—*per contra*, it as greatly obligates the disclosure of the same."[20] In his opinion, the penalties aimed at the abortionists had hardly any deterrent effect. Rather, the "party inciting the act" (that is, the pregnant woman or her husband or partner) had to be made to "fear the consequences."[21] Medical evidence in court about the abortion was therefore crucial.

However, the legal position of doctors who were asked to give evidence in abortion trials varied between the states. In some states (California, Idaho, Minnesota, Montana, North Dakota, Oregon, South Dakota, Utah, and Washington), the medical privilege statutes pertained only to *civil* actions.[22] Therefore, medical witnesses in these states had no legal grounds to refuse testimony in *criminal* proceedings for abortion. The situation was less clear in those states that had a statutory medical privilege but made no distinction as to whether it applied to both civil and criminal actions or only to civil cases. For example, in a New York trial against an abortionist, the court of appeals ruled in 1886 that the evidence of a physician who had examined the woman after the abortion was inadmissible, as his disclosure "tended to convict her too of crime or to cast discredit and disgrace upon her."[23] New York had introduced a medical privilege in its Revised Statutes of 1828 but had then incorporated it into the Code of *Civil* Procedure in 1877, whereupon the provision in the Revised Statutes had been repealed. In the state of New York, the rules of evidence for civil cases were, however, also applicable to criminal cases in general.[24] So, in this criminal case, the medical privilege, which was meant to protect the patient, obviously overrode the interest in prosecution. The physician concerned had been sent by the public prosecutor to examine the woman, but the court held that a confidential doctor–patient relationship had been established because he had also prescribed for her. Moreover, the woman had not waived medical confidentiality but on the contrary insisted upon it.[25]

By contrast, in a case of murder (by poisoning) in 1880, the New York State Court of Appeals rejected the defense's attempt to exclude the evidence of the victim's physician. The court argued that the medical privilege existed for the protection of the patient, not to shield someone who was charged with murder.[26] Also, in 1839, the New York Supreme Court ruled that a physician who had been consulted about the means for procuring an abortion by a person who later became the defendant in an action for seduction was competent to testify to the information he had obtained on this occasion. In the court's opinion, this information was not protected by the medical privilege statute.[27]

The difficulties with confidentiality in abortion cases became particularly obvious in 1889 in a New York trial of a man who had assisted his lover in a procedure from which she subsequently died. The woman, about three months pregnant, had used a wire to insert a catheter into her womb, and he had then blown into the catheter. When she had a fit or fainted, he hurried to a physician's office, asking the doctor for help and telling him what they had done so that the doctor would take the right medical equipment with him. When the case was first heard, the physician's evidence about this incident was admitted, and the defendant was found guilty of manslaughter in the first degree. However, on appeal, the New York Supreme Court reversed the verdict, holding it to have been an error to admit the doctor's testimony. The medical privilege statute, argued the higher court, "both in letter and spirit," protected the confidence that had been put in the physician and forbade him to betray it.[28]

As these cases indicate, doctors faced considerable uncertainty regarding whether they should report illegal abortions and give evidence as witnesses in abortion trials. It was difficult to draw any firm general conclusions from those precedents. Moreover, the problem was aggravated by differing medical privilege regulations among the various state jurisdictions. Following the arrest of a Massachusetts physician on the charge of "being an accessory after the fact to a criminal abortion" in 1899, the *Journal of the American Medical Association* complained about those differences from state to state: "In some localities it may come to be perilous to treat [abortion] cases as they occur without reporting all suspicious appearing facts to the authorities, while a mile or two distant such revelation would bring liability to heavy damages if not a criminal prosecution. The Massachusetts physician referred to may or may not have been cognizant of a criminal act and guilty of concealing it, but in New York he could not have done otherwise than keep this knowledge to himself."[29] In contrast to New York, the state of Massachusetts had no medical privilege statute.

Expert advice on the matter was accordingly cautious. One medico-legal author, William C. Woodward, suggested in 1902 that doctors should state abortion cases hypothetically, without disclosing the interested parties, to an appropriate law officer in order to obtain specific advice. If the advice was that they were not bound to disclose, they should keep silent. They should not approach a law officer for advice, however, if they were not prepared to give evidence when told that they had a duty to disclose.[30] Legal author William C. Tait immediately contradicted Woodward's suggestion. In his view, the legal questions concerning privileged communications were still too unsettled for any attorney to give reliable advice in such cases. Therefore, the safest route was for doctors "to obey the injunction of the Hippocratic oath" and to keep the medical information secret unless the patient had waived confidentiality or a judge had ordered them to testify.[31]

Opinions on how physicians should behave when they became aware of an illegal abortion also clashed at a medico-legal symposium organized by the Denver and Arapahoe Medical Society in 1903. Under Colorado law at that time, abortionists were punishable with imprisonment up to three years and a fine of up to $1,000; if the woman died from the procedure, the abortionist would face a murder charge. The woman herself was not punishable. Colorado also had a medical privilege statute, which said that a physician or surgeon "shall not without the consent of the patient, be examined as to any information acquired in attending the patient, which was necessary to enable him to prescribe or act for the patient." In this situation, one attorney speaking at the symposium, Allen H. Seaman, advised that doctors called to attend to a woman after an abortion should restrict their questions to only what was necessary to know for her treatment and avoid any questions about who performed the intervention and why or regarding anything else that might implicate any person in a violation of the law. The patient should be told that she needs to respond only to the questions put to her by the doctor. If questioned by the police, the doctor should keep silent on the grounds of his professional ethical duty of secrecy. He should only speak about the case if the patient had allowed it or if he was directed to do so in court.[32]

By contrast, another legal commentator held that because doctors already had a statutory obligation to report contagious diseases such as smallpox and diphtheria, a statute should be enacted to fight the "evil" of abortion by requiring physicians to report to the authorities any such case that came under their observation.[33] In response, two doctors assured their profession's commitment against the widespread practice of abortion. One of them hoped that educational measures, campaigns against advertisements for abortifacients, more effective law enforcement, and better maternity hospitals would bring about a change in public attitudes, but neither doctor addressed the suggested reporting of cases.[34] In a final statement, attorney H. H. Hawkins pointed out that the existing abortion laws in Colorado were largely ineffective: only five abortionists had been sent to prison in the last thirty-four years, and the city of Denver "swarmed" with abortionists. Apparently, no one had been prosecuted for abortion in Colorado unless the woman had died. While the fault lay, in his opinion, with public sentiment, which was too lenient toward abortionists, the physicians should act by policing their profession. Moreover, amendments to the law should permanently bar from practice any physician convicted of having performed an abortion and should require doctors to report abortion cases to the authorities.[35]

The topic of reporting abortions was brought up again at the 1909 Annual Meeting of the American Medical Association. Addressing the section on preventive medicine and public health, Philadelphia physician Myer Solis Cohen proposed that physicians should be required by law to notify the health authorities of abortions as well as accidental miscarriages. The health officer would then involve the coroner or the prosecuting attorney as appropriate—that is, if the woman had died or if a criminal abortion was suspected. Cohen saw three advantages to such a system: the improvement of vital statistics, especially in view of decreasing birth rates; the protection of the physician from charges of having committed an illegal operation; and the education of the public, who would learn that abortion was a serious matter. In fact, he claimed, some physicians already reported cases of abortion, though some only if the woman was dying or if they assumed that a criminal act had been committed. In the discussion, however, doctors from various states expressed their doubts that such legislation would be successful, given the existing difficulties with the registration of births and the reporting of tuberculosis and other notifiable contagious diseases.[36]

Unsurprisingly in this climate, reporting abortion requests to the authorities, before any intervention had taken place, was still not an option for most doctors, as New York physician Norman Barnesby pointed out in 1910 in his critique of the profession, *Medical Chaos and Crime*. In fact, he had not heard of anyone who had done this.[37] At this time, only one state, Missouri, had introduced a clause in their medical privilege statute that allowed physicians to testify to what they had learned when attending to a woman after an abortion had been performed.[38] In 1913, the Supreme Court of Wisconsin ruled that the medical privilege in court, granted in the Wisconsin statutes of 1911, was invalidated in criminal proceedings for abortion.[39]

As these examples indicate, there were three major obstacles to the proposals of American antiabortion campaigners around 1900 that physicians should be compelled to report and to give evidence about illegal abortions: (1) doctors' commitment to their traditional and professional duty of secrecy; (2) the provisions, in many states, of a medical privilege in court that protected the patient's confidence; and (3) the persisting attitudes of the public, which tended to condone early terminations of pregnancies, regardless of the fact that abortions at any stage of gestation had been made punishable in criminal law. However, as Leslie J. Reagan has shown in her study on the history of abortion in the United States, some physicians in the early twentieth century became very concerned about a possible loss of their reputations and potential prosecution as accomplices when they became involved in a case of criminal abortion. They therefore threatened the woman that they would not treat her until she revealed the name of the abortionist. Moreover, some hospitals—for example, in Chicago—collaborated with police in obtaining deathbed declarations from women about the identity of the abortionist and the procedure used.[40]

"A Monstrous Cruelty": The *Kitson v. Playfair* Trial in London and Its Impact

In late March 1896, a libel action of Mrs. Linda Kitson against her brother-in-law, Dr. William Smoult Playfair (1836–1903), an eminent London gynecologist and obstetrician, and his wife was heard at the High Court's Queen's Bench Division before Justice Henry Hawkins. About two years earlier, in February 1894, Playfair had examined Mrs. Kitson together with her usual doctor, Muzio Williams, for an unclear gynecological problem. Finding placental tissue in her womb, he concluded that she must have recently had a miscarriage—a shocking result for Playfair, as he knew she had not seen her husband, Arthur Kitson, for well over a year. Linda Kitson had arrived in London from Australia, where she had lived with her husband and two daughters, in December 1892, while her husband had stayed behind to deal with the various problems of his failing business. In Playfair's view, it was obvious that she must have been adulterous, though he was aware that she had had a miscarriage shortly before she travelled to England. In a subsequent correspondence, he put before her the alternative either to leave London, thus breaking off all contact with his family, or to give him assurances that her husband had secretly visited England in the last three months; otherwise, family honor would demand that he inform his wife, Arthur Kitson's sister. After Linda Kitson refused to leave and remained evasive about a possible recent visit of her husband to London, Playfair disclosed to his wife what he had found during the physical examination of his sister-in-law and the conclusion drawn from it. At his wife's instigation, he then also informed Sir James Kitson, his wife's older brother. As the head of the Kitson family, which had become wealthy in the Leeds iron industry, Sir James had paid an annual allowance of £400 to Linda to support her and her daughters in England while her husband was away in Australia. On receiving the news about her apparent adultery, he stopped the allowance. Linda Kitson subsequently sued Playfair and his wife for slander and libel based on the disclosure that they had made to Sir James Kitson. The trial, which was reported in great detail in *The Times* as well as *The Lancet* and the *British Medical Journal*, ended with a verdict against Dr. Playfair. The jury concluded that Playfair's communication to Sir James had not been made "in good faith and without malice, or from a sense of duty, but from an indirect motive." Linda Kitson was awarded £12,000 in damages, a very high sum. Following an appeal, the parties eventually agreed on the reduced amount of £9,200.[41]

As Angus McLaren has argued in his historical analysis of this case, the trial's outcome owed much to gender and class issues in late Victorian society: the jury showed sympathy towards Linda Kitson as a distressed, middle-class lady who had been accused of immorality by a relentless male relative who failed to act like a gentleman toward her.[42] In the contemporary discussions of this case and the trial itself, however, two further aspects were prominent: the question of justified breaches of medical confidentiality and whether doctors should report abortion cases to the authorities (although Linda Kitson had not been accused of having had an illegal abortion).

These topics of debate largely arose because the defense team, led by Sir Frank Lockwood QC, did not follow the strategy of "justification" (that is, to show that the allegation of adultery was true) but argued that Dr. Playfair's communications with his wife and his brother-in-law were "privileged" communications arising from a duty to his family and therefore immune to charges of slander and libel. In this context, expert witnesses were called to testify not only to the medical aspects of the case but also to the professional conventions ruling the confidentiality of the doctor–patient relationship. Cross-examined on the latter topic, Sir John Williams, like Playfair a leading obstetric physician, stated that there were three exceptions to the general rule of keeping patient confidences secret. Disclosure was justified (1) when the medical evidence was required in a court of law; (2) in order to inform the public prosecutor about an intended or committed crime; and (3) to protect the doctor's wife and children.[43] Sir William Broadbent, senior censor of the Royal College of Physicians, who was next called as a witness, confirmed these three exceptions to medical confidentiality but was noticeably hesitant about the last one, admitting that disclosure in this situation should only be made as a last resort.[44] Obviously, Williams's third exception could have potentially exonerated Playfair. Particularly remarkable, however, was Justice Hawkins's intervention during Williams's cross-examination. After Williams confirmed to the plaintiff's counsel that it was for the doctor to decide under which circumstances he was obliged to breach confidentiality, Hawkins put a hypothetical case to the witness: "Suppose a medical man were called in to attend a woman, and in the course of his professional attendance he discovers that she has attempted to procure an abortion. That being a crime under the law, would it be his duty to go and tell the Public Prosecutor?"[45] Williams replied that the Royal College of Physicians had answered yes to this very question, to which Hawkins sarcastically remarked, "Then all I can say is that it will make me very chary in the selection of my medical man."[46]

Moreover, in his summing up statement at the end of the trial, Hawkins returned to the issue of medical confidentiality. Not only did he make the point here that it was for the judge to decide whether a doctor had to give evidence about his patient, perhaps allowing a doctor to keep silent under particular circumstances (an important comment on the question of a medical privilege in court;

see chapter 1), but Hawkins also came back to his example of a woman needing medical aid after an abortion attempt. It would be "a monstrous cruelty," he said, if the doctor then reported her to the public prosecutor or the police.[47] The fact that Hawkins, as a judge, publicly endorsed not reporting the "crime" of abortion under certain circumstances would become significant for the medico-legal debate on this subject.[48]

The immediate comments in the lay and medical press focused, however, on the questions of whether Playfair's breach of confidentiality was morally justifiable under the given circumstances and what lessons could be learned from the trial for medical secrecy in general. An editorial in *The Times* concluded that Playfair had made "a grave mistake of judgment" in believing that he had to disclose Linda Kitson's condition to his wife and Sir James Kitson. The general lesson from his case was that "if a medical man reveals a professionally gained secret he does so at his peril." Comparing the patient's confidence with a confession to a priest, the editorial held that doctors should keep strict confidentiality, the only exception being when asked to give evidence in a court of law. Even in this situation, doctors should stay silent in order to preserve their patients' trust in them.[49] Several letters to the editor of *The Times* further problematized the issue of medical confidentiality, with opinions ranging from a call for an absolute duty of secrecy for doctors, also in court, to the suggestion of "some readjustment" of the views on privilege in order to protect the community from crime, vice, and disease.[50]

For the editors of *The Lancet*, it was a "painful duty to assent to the proposition that Dr. Playfair did not act as discreetly as he might have done." Confidentiality should be breached only for "overpowering reasons," and the medical practitioner should have weighty information to sustain the allegation of a crime such as abortion before reporting the case.[51] When, in a subsequent letter, a "Young Practitioner" asked whether he should report to the police a young, unmarried woman who had admitted to him an illegal abortion attempt but was now recovering, *The Lancet*'s editors replied with "no"—they did not think that it was a medical man's duty to inform the authorities under such circumstances.[52] A further *Lancet* article commented on a recent divorce trial in which the medical witnesses had tried to refuse giving evidence (with reference to the *Kitson v. Playfair* case) until they were compelled by the bench to testify. In light of this, "an authoritative statement of the still unpublished law of professional silence" would be welcomed by medical practitioners.[53]

The *British Medical Journal*, by contrast, displayed a sympathetic attitude toward Playfair, accepting that he had been motivated by "a strong sense of family duty"[54] and acknowledging that he had experienced an exceptional moral conflict: "Never before, so far as we know, has the tradition of medical secrecy been so severely tested. Dr. Playfair had to balance the grand traditions of medical confidence as against duties between relative and relative. We find it difficult to say how any other man placed in a similar position would have acted, and we have yet to learn that those who censure him so freely would not have taken the same course."[55]

The *BMJ* believed that the *Kitson v. Playfair* trial had actually strengthened the doctrine of medical secrecy, albeit at the cost of one individual doctor.[56] While this claim may have been overstated, the case certainly brought the topic of confidentiality to the forefront, as several subsequent letters to the editor showed. Some correspondents focused on the medical aspects of the case—that is, for how long after a miscarriage placental tissue could be retained in the uterus and the morphological changes it would undergo during this time.[57] But others commented on the trial's implications for the understanding of medical confidentiality. A legal correspondent, for example, took up Justice Hawkins's question of whether a doctor should report a patient who had undergone an illegal abortion. As the commentator pointed out, Hawkins's view that the doctor was not necessarily entitled to do so was legally correct, because the secret was the patient's, not the doctor's, and the woman's confidence rested on the implied understanding that her doctor would use the information in her interests only. The matter would be different if the abortion was still to be carried out. If the doctor kept his knowledge about this plan to himself and refrained from informing the authorities, he might make himself an "accessory to the crime" or "practically allow the crime to be committed."[58] Another correspondent, signing M.D., emphasized that Justice Hawkins's characterization of reporting an abortion attempt to the public prosecutor or the police as "a monstrous cruelty" had only the status of an *obiter dictum*—that is, an opinion expressed from the bench that was not binding to other judges. Consequently, Hawkins's opinion could not be accepted as a guide to conduct.[59] Other commentators took issue with the breaches of confidentiality required of doctors in English courts. As one of them noted, in England, where doctors were frequently compelled to disclose their patients' confidences in divorce trials, medical confidentiality could not be held inviolate as it was in other countries, and many other reasons justifying disclosure had to be admitted.[60] In two further editorials, the *BMJ* tried to clarify the issues surrounding privilege and confidentiality and to propose general rules of conduct.[61] One such rule was not to disclose patients' secrets without their consent, except if this was demanded by the law. More specifically, if a crime had already been committed and could no longer be undone, the doctor was entitled to keep the secret if the (guilty) patient refused to permit disclosure. However, if the crime could still be prevented by the doctor's disclosure, he was justified in revealing the patient's secret; otherwise, he would make himself an accomplice in the crime. With regard to illegal abortion, for example, "where [. . .] a doctor believes himself to have evidence that some quack is making a custom of nefarious practices," it was the doctor's duty to inform the police.[62] Ultimately, however, the *BMJ* was reluctant to make authoritative statements in these matters, calling instead, like *The Lancet*, for the professional bodies to do this.[63]

Legal Advice and the Views of Medical Professional Organizations in Britain

In November 1895, before the case of *Kitson v. Playfair* came to court, the Royal College of Physicians (RCP) of London had already decided to form a committee "to define in a legal sense the proper conduct of a practitioner when brought into relation with a case of acknowledged or suspected criminal abortion, with power to take legal advice."[64] Among the committee members were two of the expert witnesses in the subsequent *Playfair* trial: Sir William Broadbent and Sir John Williams. In England, performing an abortion, including self-abortion by the pregnant woman, was then punishable under the Offences against the Person Act of 1861, which threatened imprisonment for life as the maximum penalty. However, in practice, prosecutions under this act were likely to be carried out only if the woman had died, and the law does not seem to have had a significant deterrent effect.[65] Still, the status of abortion as a crime, unless it was performed by a doctor to save a woman's life, created considerable uncertainties for medical practitioners. Involvement in an abortion case could damage their reputations and lead to accusations of having performed an illegal operation. By the end of April 1896, the RCP committee had obtained legal advice from Edward Clarke QC (1841–1931), a former solicitor general, and his colleague Horace Avory (1851–1935), who later became a High Court judge.[66]

As the committee report detailed, Clarke and Avory advised that the law did not forbid procurement of an abortion or even destruction of the fetus during labor if this was necessary to save the mother's life. If a medical practitioner knew or believed that he was attending a woman on whom a criminal abortion had been performed, he should treat her to the best of his skill. He thereby did not make himself liable as an accessory after the fact, as long as he did not do anything to assist the patient in escaping from justice. He also could not be charged with misprision of felony (that is, concealment of a crime) if he did not inform the authorities of a merely suspected illegal abortion. If the abortion had not yet taken place and the pregnant woman had given her doctor the name of the person who would perform the procedure, he should immediately warn the person concerned that he had been told about this plan (with the apparent intention to deter the abortionist from carrying out the intervention). In general, Clarke and Avory concluded, the medical practitioner should exercise his discretion as to when to report a particular case or what to do if the pregnancy endangered the woman's life.[67]

Though originally a confidential document, the Royal College's committee report was subsequently published, against the wishes of the college's Censors' Board, in the journals *The Scalpel* and *Medical Press and Circular*.[68] In this way, the question of what a medical practitioner should do if he became involved in an abortion case received some public exposure in Britain, in addition to the discussions that arose from the *Kitson v. Playfair* trial. In his guidebook *Medical Ethics* (1907), Robert Saundby, the vice president of the British Medical Association as well as a fellow of the Royal College of Physicians and a member of the General Medical Council, referred to the case of *Kitson v. Playfair* as well as to Clarke's and Avory's legal advice to the RCP. Somewhat overstating their advice, he suggested that the two lawyers' opinion "was to the effect that a medical man should not reveal facts which had come to his knowledge in the course of his professional duties, even in so extreme a case where there were grounds to suspect that a criminal offence had been committed."[69] Professional opinion, both medical and legal, at this stage of the discussion seemed to favor preserving confidentiality over prosecution. There was, however, no clear-cut guidance, and the decision was left to the individual doctor under the specific circumstances of a case.

This is illustrated by an abortion case near Liverpool in 1912, which led to a coroner's inquest, as the woman had died from septic peritonitis after the intervention. The doctor who had treated her gave evidence about her case at the inquest, but the patient had refused to divulge the abortionist's name, and there was no clue to the latter's identity. A police superintendent suggested at the inquest that the doctor should have reported the case before the patient's death. The medical practitioner concerned, Dr. R. Hill Brown, replied that he had been bound to confidentiality and that he had the duty to do the best for his patient—a position that was supported by the coroner.[70]

In 1913, the Local Government Board asked the censors' board of the RCP for advice on such cases. The inquiry was due to the conflicts doctors frequently experienced in public hospitals when they had to treat patients whom they suspected of having had criminal abortions. On the one hand, it could be said that, as medical officers in a public institution, they had a duty to help in detecting criminal offences. On the other hand, the outcome of the *Kitson v. Playfair* trial was still in doctors' memories, and Saundby's book appeared to discourage the reporting of such cases. The college responded by releasing their 1896 report, including Clarke's and Avory's advice, to the Local Government Board.[71] Shortly afterward, however, in 1914–15, the issue again became prominent.

Horace Avory, who had meanwhile been appointed as a judge, attracted attention with an address to the grand jury at the Birmingham Assizes in December 1914 regarding charges against a suspected abortionist, Annie Hodgkiss. In October of that year, a thirty-two-year-old single woman, Ellen Agnes Armstrong, had died from septicemia (blood poisoning) after her pregnancy had been terminated in about the fifth month. Following the intervention, when she had a stillbirth and then fever and vomiting set in, a district nurse and three doctors successively treated her, initially at her aunt's home and then in the Women's Hospital at Sparkhill, to which she had been referred by her usual medical practitioner, Dr. Arthur William Aldridge. Deeply ashamed of her situation—the man who had gotten her pregnant ignored her—Ellen Armstrong had been determined to have an abortion: "She would rather die than have the disgrace," as she had told her father when he visited her after the intervention. She had refused to give her father or her aunt the name of the woman who had performed the abortion. Eventually, she had confided the name and address of the abortionist to Dr. Aldridge, but neither he nor any of the other doctors involved had informed the police or the legal authorities before she died. Only after her death did the surgeon who had treated her in the hospital, John Fermeaux Jordan, report the case to the police.[72]

While the coroner's inquest returned a verdict of willful murder, Justice Avory felt he was unable to convict the suspected abortionist, Annie Hodgkiss, because no statement had been obtained from Ellen Armstrong that could be used in court. The only written piece of evidence was a note sent by Hodgkiss to Armstrong in which she confirmed she would see her at a date in early September.[73] Avory concluded,

> I can see no evidence, as the case now stands, which will justify you in finding a true bill against the prisoner for murder. The law provides that in the case of any person who is seriously ill, and who, in the opinion of a medical man, is not likely to recover, the evidence of such a person may be taken by a Justice of the Peace. Under circumstances like those in the present case, I cannot doubt that it is the duty of the medical man to communicate with the Police, or with the Authorities, in order that one or other of those steps may be taken for the purpose of assisting in the administration of justice. [. . .] It may be the moral duty of the medical man, even in cases where the patient is not dying, or not unlikely to recover, to communicate with the Authorities when he sees good reason to believe that a criminal offence has been committed. However that may be, I cannot doubt that in such a case as the present, where the woman was, in the opinion of the medical man, likely to die, and therefore, her evidence was likely to be lost, it was his duty, and that some one of those gentlemen ought to have done it in this case.[74]

Moreover, Avory noted that the advice he had given some twenty years ago to the Royal College of Physicians had been "either misunderstood or misrepresented in a text-book of medical ethics" (that is, Saundby's *Medical Ethics*).[75] In essence, he thus blamed the doctors involved in Ellen Armstrong's case for the failure to obtain evidence that could have led to the conviction of a dangerous abortionist.

Avory's comments led to strong reactions from the organized medical profession after public prosecutor Sir Charles Mathews brought them to the attention of the Royal College of Physicians as well as the British Medical Association and the Royal College of Surgeons.[76] The BMA's solicitor, William Hempson, advised that there was no legal obligation for a doctor to inform the police if, during treatment, he had come to know about a criminal act committed by his patient. He also read Clarke's and Avory's 1896 advice to the RCP in that way, even in light of Avory's recent comments.[77] Following a discussion in its central ethical committee and a meeting of delegates with the Lord Chief Justice, the attorney general, and the public prosecutor, the BMA's council confirmed, in July 1915, its resolution that (1) "a medical practitioner should not under any circumstances disclose voluntarily, without the patient's consent, information which he has obtained from that patient in the exercise of his professional duties" and that (2) "the State has no right to claim that an obligation rests upon a medical practitioner to disclose voluntarily information which he has obtained in the exercise of his professional duties."[78]

The Royal College of Physicians, after discussing the matter with the public prosecutor, resolved that a medical practitioner was not justified to disclose a patient's information without her consent. Instead, he should urge the woman, especially if she was dying after the criminal abortion, to make a statement that could be used as evidence against the abortionist. If she refused, however, the doctor would be under no legal obligation to pursue the matter further; he should continue to attend to the patient to the best of his ability. Before taking any action, if the patient did make a statement, the doctor was advised to seek legal advice to ensure that the patient's statement was valid and to protect himself against potential subsequent litigation. If the patient died, he should refuse to fill in the certificate of death and instead inform the coroner.[79] The Royal College of Surgeons also rejected Avory's views about an obligation to report cases of criminal abortion and wished to leave the matter up to individual medical practitioners.[80] *The Lancet* declared in an editorial, "The medical man is there to save the patient's life and to restore the patient to health if he can; not to aid the police by playing the detective. [. . .] If he once reveals his patient's secret, with, be it remembered, the practical certainty that his doing so will become publicly known, he will destroy belief in his professional discretion [. . .] and any general adoption of his course of action would diminish or destroy all trust in his profession everywhere."[81] In their view, the patient would then conceal her illness, with fatal consequences, or she would seek unqualified treatment, perhaps even from the person who performed the abortion.[82] Moreover, *The Lancet* pointed out that Justice Hawkins's remarks during the *Kitson v. Playfair* trial were at variance with Avory's views.[83] The judiciary therefore appeared to be inconsistent in this matter. The journal's advice to medical practitioners was to consult a fellow practitioner if they found themselves in charge of a case of abortion.[84]

The British medical profession thus took in general a defensive position against legal demands to sacrifice confidentiality in the interest of prosecuting abortionists. Clearly, abortion was one of those situations where patients expected secrecy from their doctors, and disclosure under these circumstances would have damaged doctors' reputations and practices. They might also have to face charges of slander and libel. Public expectations of medical secrecy were powerfully conveyed, for example, through the new medium of film. A 1930 motion picture by Edgar Wallace and Manning Haynes titled *Should a Doctor Tell?* featured the character of a London consultant who steadfastly refused to give evidence during a divorce trial, accepting a heavy fine for contempt of court. But then he finds that his son's fiancée is the same young woman who had become pregnant in a relationship with a married man and had sought his medical help several months earlier. Her child having meanwhile "died," she implores him: "I want your silence. I don't plead with you as a father, I demand it of you as a Doctor!"[85]

As the real, tragic case of Ellen Armstrong had demonstrated, only the death of the woman concerned would give doctors sufficient reason to report an abortion case to the authorities.[86] British medical attitudes toward this matter thus resembled those prevalent among American doctors. However, lawyers' views in Britain as well as the United States were not unanimous on the question of whether doctors had a duty to report cases in which the woman was recovering from a criminal abortion or whether their professional obligation of confidentiality should be respected in this situation. The even further-reaching suggestion, discussed in the United States, that doctors should inform authorities if a request for an abortion had been made to them, was also addressed in Britain in the aftermath of the *Kitson v. Playfair* trial. While in America, the existence of a medical privilege in many states provided an argument against disclosing abortion cases, in Britain, the medical professional organizations resisted proposals of such notification.

German Comments on Confidentiality in Abortion Cases

In imperial Germany, abortion was punishable with penal servitude of up to five years or, in case of mitigating circumstances, an imprisonment of no less than six months under section 218 of the Penal Code of 1871. These penalties applied to women who had performed an abortion on themselves as well as to abortionists and pertained to terminations of pregnancy in any period of gestation. Though not explicitly regulated by law, only abortions carried out by a doctor in order to save a woman from an acute danger to her life were legally accepted as maternal acts of self-defense against the unborn child. As in the United States and Britain, however, the status of abortion as a criminal offence had little deterrent effect. It was estimated that three hundred thousand to five hundred thousand abortions were carried out annually in Germany around 1900 and that of the 20 percent of all pregnancies that ended in miscarriage, about half had been criminally induced. The relatively harsh abortion law contrasted with the prevailing attitudes of the German population, many of whom regarded terminations of pregnancy before quickening as morally acceptable. Only a small portion of all abortion cases ended with a criminal conviction—between 1882 and 1912, fewer than a thousand per year.[87]

In this situation, German legal and medical authors, as in Britain and the United States, discussed whether doctors had a duty to inform the authorities if they had become aware of a case of illegal abortion. Under section 139 of the Reich Penal Code, breaches of professional secrecy were permitted if they could assist in preventing a serious crime (see chapter 1). However, it was questionable whether this exception applied to criminal abortion. Moreover, German doctors had a right to refuse to give evidence in court on the basis of section 52 of the Code of Criminal Procedure, but it was unclear whether this meant they should keep silent when called as witnesses in trials for illegal abortion. In 1894, several medical practitioners in the county of Lennep attracted attention by their unwillingness to breach confidentiality when they were called as witnesses in an inquest for criminal abortion.[88] In the opinion of their district physician, Dr. Schlegtendal, section 300 of the Penal Code guaranteed medical secrecy for "the girl or the married woman who comes to me for treatment with an imminent miscarriage or with a septic infection and in whom I detect *with* my questions or *without* questions the most significant signs of a criminal intervention against the developing life."[89] Not everyone agreed with this view, however.

In his book on medical confidentiality (1st ed., 1893), the Berlin medico-legal expert and neurologist Siegfried Placzek, for example, claimed that the majority of physicians would report cases of criminal abortion to the police, regardless of their duty of professional secrecy and even though they were not legally obliged to inform on already committed crimes. In Placzek's view, reporting a criminal act to the relevant authorities could never be illegal. Still, he thought that the decision of whether to report a case should be left to the discretion of the doctor concerned, as patients' trust would disappear if a medical practitioner turned into an informer.[90]

In his textbook on medical ethics (1902), Placzek's Berlin colleague Albert Moll advocated a more lenient stance toward women who had illegal abortions. Emphasizing that abortion was, in general, regarded as less reprehensible than other criminal acts and that its moral significance was controversial from an ethical perspective, he doubted that a doctor should report it, especially as this would expose his patient to a conviction. Punishing a person for a criminal offence that many did not recognize as a danger to the public was, in his opinion, an insufficient reason to breach the duty of medical secrecy.[91] As Moll observed, early abortion after rape, or if a woman already had many children, was seen as morally acceptable by many people, and there was a remarkable discrepancy between public attitudes and the severity of the penal law.[92] Yet Placzek, developing his discussion of the topic in the third edition of his book (1909), argued that physicians were justified to inform on someone who had made an abortion request—in order to prevent the "crime."[93] Moreover, if the woman died as a result of the illegal procedure, the abortionist should be reported to the authorities.[94]

The issue of reporting abortion cases was also addressed in legal studies considering the implications of section 300 of the Reich Penal Code. There was consensus that under the Reich laws doctors had no obligation to inform the authorities on an already performed illegal abortion, although such a duty had been stipulated in ordinances of some German states (e.g., in Baden in 1883 and Saxony in 1892 and for midwives in Bavaria in 1899).[95] Furthermore, a doctor was not thought to be legally entitled to breach professional secrecy in order to prevent a planned abortion.[96] He could, however, report a professional abortionist to the police if he had appropriate evidence.[97] Regardless of such legal clarifications and ethical advice, doctors faced a morally difficult situation when confronted with cases of abortion. In 1911, Moll commented on a case in which a doctor had reported the abortionist to the public prosecutor with the consent of the patient concerned. As a consequence, both the abortionist and the woman on whom the abortion had been performed were put on trial and convicted. However, the woman submitted a plea for clemency, which was supported by the public prosecutor. In Moll's view, the case was one in which the doctor had to make a conscientious personal decision, as there was no legal duty to report, but he warned that medical practitioners should not construct for themselves a "higher moral duty" to breach confidentiality on every occasion.[98]

In 1914, the German Supreme Court made an important decision regarding doctors' right to refuse giving evidence in abortion trials (under section 52 of the Code of Criminal Procedure). Two doctors had testified to several cases in which the defendant was accused of having performed an abortion, but they were not willing to reveal the names of the women concerned. Following an appeal of the defense, the Supreme Court confirmed that doctors were entitled to give incomplete evidence on the grounds of their legal duty of confidentiality and their right to refuse testimony and that such incomplete evidence was admissible if it did not lead to incorrect and untrue statements.[99] In this way, the Supreme Court allowed prosecutions of abortionists without necessarily implicating the women concerned—thus going a step further than in the court case mentioned by Moll in 1911.

As these examples show, German doctors were, like their American colleagues in some states, protected by their medical privilege in court when information on abortion cases was demanded of them. Moreover, German doctors had no duty under Reich legislation to report cases of abortion, let alone plans for an abortion. If they did report without their patients' consent, they risked charges for the breach of professional secrecy under the Penal Code. By World War I, it had become clear in Germany, as well as the United States and Britain, that the majority of doctors were unwilling to breach confidentiality in abortion cases. Only if the woman had died from complications following the intervention did they feel justified to inform the authorities, as this might protect other women from the activities of a dangerous abortionist.

Abortion, Sterilization, and Confidentiality in Germany during the 1920s and 1930s

In the 1920s in Germany, calls to reform the abortion laws became more insistent. While protests by the German women's movement had already begun in the early 1900s and discussions about mitigating the penalties for abortion had been part of the deliberations in plans for a general reform of the Reich Penal Code before World War I, abortion law reform became a highly politicized issue in the Weimar Republic. Although it was not official party policy, the German leftist parties, the Social Democratic Party (SPD), the Independent Social Democratic Party (USPD), and German Communist Party (KPD), supported women in their campaign for a repeal or radical amendment of section 218. In part, such support was motivated by the aim to gain female voters after women's suffrage had been introduced with the elections of January 1919. On the other hand, the Left acknowledged the hardships for women during times of economic crisis and to some extent accepted their use of abortion as a means to limit the size of their families. During the Weimar Republic, the number of illegal abortions was estimated to be between four hundred thousand to one million per year.[100]

Whereas the medical profession had taken a predominantly pronatalist, antiabortion stance before World War I, the political divides of the 1920s became apparent in doctors' opinions on this issue. A conservative majority of doctors, many of them part of the German Medical Association, opposed a reform of the abortion law, but Socialist physicians and some members of the Federation of German Women Doctors supported change. In 1926, section 218 of the Penal Code was revised: punishment for the aborting woman and her helper was reduced from penal servitude to imprisonment but still included penal servitude for those who procured an abortion for money or sold abortifacients. Under mitigating circumstances, the penalty for the woman who had undergone the abortion (during this time classed only as a misdemeanor) could be reduced to one day's imprisonment or a fine of three marks. A 1927 German Supreme Court decision permitted doctors to perform abortions on therapeutic grounds.[101]

Abortion policy radically changed, however, after the National Socialists had come to power in early 1933. The Nazi policy showed a characteristic dichotomy on the basis of racial and eugenic discrimination. On the one hand, the Law for the Prevention of Hereditarily Diseased Offspring (*Gesetz zur Verhütung erbkranken Nachwuchses*, July 14, 1933) allowed for compulsory sterilization and—following a memorandum from the head of the Reich Medical Chamber, Dr. Gerhard Wagner, and revision of the law in 1935—abortion for eugenic reasons. In September 1940, Wagner's successor, Leonardo Conti, the new head of the Reich Department of Health, authorized abortions on racial grounds, a measure that especially targeted Polish women in the occupied territories. On the other hand, abortions performed on German women with "valuable" or "flawless hereditary material" were again placed under severe penalties—in order to protect German "national vitality" (*Volkskraft*). Abortion services and the advertising of abortifacients were banned. A Reich-wide system of assessment centers for therapeutic abortions and sterilizations was set up by 1935, and the revision of the law on eugenics in the same year made it mandatory for midwives and attending doctors to report every miscarriage, abortion, or stillbirth to the public health authorities, including the name and address of the woman concerned. This information was used to investigate possible cases of illegal abortion. Criminal charges for abortion increased, and prosecutions were carried through with increasing severity. As Gabriele Czarnowski has documented in her study on abortion in Nazi Germany, between 1942 and 1944, twenty-five people were sentenced to death as abortionists.[102]

Besides the mandatory reporting of abortions, the Law for the Prevention of Hereditarily Diseased Offspring had other significant, more general implications for medical confidentiality. It obliged doctors to report patients with conditions such as inborn mental deficiency, schizophrenia, manic-depressive illness, epilepsy, Huntington's chorea, blindness, deafness, severe malformation, or severe alcoholism to the health authorities in order to have them assessed for compulsory sterilization. Section 7 of this law explicitly annulled the duty of medical confidentiality for such cases, requiring doctors to give evidence in the (nonpublic) proceedings of the newly created Hereditary Health Courts (*Erbgesundheitsgerichte*). The same law bound everyone involved in the assessment procedures and the sterilization operations to strict secrecy, threatening imprisonment of up to one year for indiscretions.[103] As an article in the official journal of the Berlin Medical Chamber explained, "Not the patient, but the state now releases [the doctor] in hereditary health proceedings from the duty of secrecy; because the purpose of this law, 'the elimination of hereditarily diseased offspring,' is that of a law fighting for the race [*Rassekampfgesetz*], which has to put the interest of the individual racially inferior member of the *Volk* last."[104] When in spring 1935 a doctor and owner of a private clinic refused to hand over the case history of a former patient to a Hereditary Health Court, referring to his duty of confidentiality according to section 300 of the Penal Code, he was fined 150 marks, a penalty that was confirmed on appeal.[105]

Eugenic thinking, as is well known, was also prominent in Britain and the United States during the early twentieth century. Between 1907 and 1917, sixteen American states, starting with Indiana, passed sterilization laws that targeted those with epilepsy, mental illness, and "mental defects" in state institutions and authorized compulsory sterilization in habitual criminals and sex offenders.[106] By 1934, twenty-seven American states had legislation for sterilization in effect, and although there were legal challenges at the Supreme Court level, by 1932, twelve thousand people had been sterilized under these laws.[107] In the mid-1930s, the aims and objectives of the British Eugenics Society included (voluntary) sterilization and the legalization of the termination of pregnancy.[108] In 1934, a departmental committee of the Ministry of Health recommended the legalization of voluntary sterilization of people with mental defects or disorders, people with transmissible physical disabilities such as hereditary blindness, and individuals likely to transmit mental disorders or defects. However, the British government, assuming that public support for such measures was insufficient, did not take up this proposal.[109] In contrast to National Socialist Germany, in Britain and in the United States, the legal status of medical confidentiality was not changed due to the rise of eugenics. One of the key arguments in Britain against doctors promoting eugenic agendas was that doctors were primarily responsible for the health of their individual patients rather than the public health.[110] Confidentiality was an integral part of that responsibility. This perspective was diametrically opposed to the ideology of National Socialism, where the individual's interest had to give way to collective interests.

Significantly, the German law on medical secrecy itself was changed. The duty of medical confidentiality was included in a new Reich Medical Ordinance (*Reichsärzteordnung*) in December 1935, and section 13 of this ordinance, which newly regulated medical secrecy, superseded section 300 of the Reich Penal Code as far as it concerned the medical profession. The penalty for unauthorized disclosure was increased from three months to one year of imprisonment and/or a fine, and the duty of confidentiality was extended from the doctor and his professional staff to include medical students who participated in professional practices and the surviving dependents of a deceased doctor or the executor of his estate. Most importantly, however, subsection 3 declared disclosure as *not* punishable if the secret was revealed "in order to fulfil a legal or moral duty" or "for a legitimate purpose according to the healthy sentiment of the people (*zu einem nach gesundem Volksempfinden berechtigten Zweck*)" and if the "threatened legal good" carried more weight.[111]

While a widening of the circle of individuals who were bound to confidentiality had already been discussed before World War I and during the Weimar Republic as part of various drafts of a new Reich Penal Code,[112] the third paragraph of section 13 more particularly reflected the changed circumstances under the National Socialist dictatorship. A predecessor of this paragraph can be identified in the draft Penal Code of 1927, where, according to section 325 (3), disclosure of a private secret was not punishable if it was necessary to serve a legitimate public or private interest, under the condition that there was no other way to serve this interest and if the "endangered interest" was more important.[113] This rather cautious formulation would have allowed for public health measures, such as disclosure to the health authorities and warning of contacts in certain cases of venereal disease (see chapter 2). However, when the National Socialists came to power in 1933, this draft became obsolete. The explicit purpose of the Reich Medical Ordinance of 1935 was to serve as "an instrument of National Socialist health policies," as the official leader of the German medical profession, Gerhard Wagner, explained. It bound the doctor to act to maintain and improve the "hereditary material" and race as well as the health of the German people. Doctors now had to directly support the health policies of the National Socialist Party (NSDAP) and the state by providing expert reports or by nominating experts.[114] Clearly, the third paragraph of section 13, by broadly removing obstacles to disclosure, served these new aims of the Nazi state.[115]

This was likewise reflected in expert commentaries and academic studies on the new regulations for medical confidentiality. Albert Hellwig, as District Court director in Potsdam, assumed that the clause about legitimate purposes according to the people's sentiment could pertain to cases in which a doctor, without having a legal duty to do so, aided in the arrest of a criminal.[116] (He did not mention, though, that one type of criminal might be an abortionist.) His Berlin colleague, District Court Director Wilhelm Schmitz, interpreted the same passage in the sense that a doctor would now be permitted to breach medical confidentiality in order to actively report a serious crime to the police—not just make a disclosure during his statement as a witness for the police or in court.[117] Wolfgang Mittermaier (1867–1956), a professor of law at the University of Heidelberg, similarly thought that the new regulation entitled the doctor to give evidence if his information was crucial for elucidating a criminal case that was more important than keeping the patient's secret.[118] A Marburg University law thesis that discussed the implications of section 13 of the Reich Medical Ordinance in detail shared this view. Committed to the National Socialist ideology, its author moreover argued that doctors, due to their special role in improving "hereditary and racial purity," were also expected to breach confidentiality and disclose relevant illnesses in proceedings for the annulment of marriages on the basis of the marital health law (*Ehegesundheitsgesetz*, October 18, 1935).[119] Another law thesis, this time from the University of Innsbruck, focused on the implications for the social insurance sector. Its author concluded that the clause about the "healthy sentiment of the people" would give the doctor a right to disclose information regardless of the patient's insistence on confidentiality if his statement would lay open the true circumstances of a case in which the patient was actually not entitled to receive health insurance benefits (that is, in order to detect insurance fraud). The logic of this suggestion derived from the National Socialist principle, highlighted in the thesis, that any protection of the individual had to stop at the point where it became an obstacle to the interest of the *Volk* or the state.[120] In addition various commentators agreed that the clause would allow a doctor to reveal patient secrets in court in an action to collect his fees or if it was necessary for his defense in a malpractice suit.[121]

As these examples illustrate, the new regulations, while ostensibly strengthening confidentiality by threatening higher penalties and widening the circle of persons obliged to it, actually considerably weakened it by permitting disclosure whenever this served the interests of the National Socialist state. Still, the specific question of a doctor's duty to report illegal abortions remained controversial, regardless of the fact that such reporting had become mandatory in 1935 with the revision of the Law for the Prevention of Hereditarily Diseased Offspring. Mittermaier still held in 1936 that a doctor should, in principle, keep the patient's secret if he had accidentally found signs of an abortion during a physical examination for another purpose.[122] The author of a medical thesis at the University of Marburg in 1939 took the same view. Aware of the legal duty to report any miscarriage before the thirty-second week of pregnancy to the health authorities, he—somewhat naively—suggested that it was not necessary to write down that it had been caused by a criminal intervention. Only if this occurred more frequently should the doctor fully report the information, because then the "threatened legal good"—the state's claim to children—carried more weight. Conforming with Nazi ideology, the author held that it was in line with the "healthy sentiment of the people" referred to in section 13 of the Reich Medical Ordinance to breach confidentiality if the woman had died after the abortion, in order to stop the abortionist "who robs the German *Volk* of its offspring and endangers the health of future mothers." In general, though, he held that patients' confidence should be protected, because otherwise they might seek help from quacks.[123]

Other authors, by contrast, proposed extending the duty to report to those cases where there was merely a suspicion that an abortion had taken place or that an abortion might be sought.[124] The author of a 1938 law thesis at the University of Erlangen suggested that if a doctor, while treating a pregnant woman, heard that she had been offered a termination by an abortionist, he was entitled according to the "*Volk*'s healthy sentiment" to report this fact to the police and public prosecutor. If the abortion had already been performed, the doctor should consider the higher legal good. The author thought, however, that the people's and the state's interest in prosecution was in this case more important than the interest in secrecy.[125] More generally, he asserted that the question of disclosure had to be decided according to the principles of the National Socialist worldview, which was "the highest and purest expression" of the people's sentiment.[126] A Rostock law thesis in 1940 took the same view, arguing that in his decision on disclosure, the doctor had to "free himself of all subjective opinions" and to follow the "sentiment of the racially and ethnically healthy parts of the *Volk*," which was "reflected in National Socialism." Avoiding the issue of abortion, he maintained that a breach of confidentiality was justified in serious crimes, such as murder or sexual crimes, when the perpetrator posed a danger to the public and there was a risk of reoffending. Yet the doctor was not supposed to "play the role of a policeman or of the public prosecutor."[127] However, as the author of a Würzburg medical thesis plainly expressed it in the same year, it was the duty of a National Socialist doctor to report illegal abortions to the authorities, because he had to put the interests of the *Volk* before the individual interest in confidentiality.[128] Berlin District Court Director Wilhelm Schmitz even went a step further by suggesting that the doctor was allowed to breach confidentiality to inform the family or the health authorities in cases of unmarried pregnant women, regardless of their wish to keep their situation secret, in order to prevent infanticide or abortion. However, Gustav Aschaffenburg (1866–1944), a prominent forensic psychiatrist from Cologne, sharply contradicted this proposal by arguing that such disclosure would seriously damage the doctor–patient relationship, preventing women in this situation from consulting a doctor, which was not in the public interest.[129]

Section 13 of the German Reich Medical Ordinance remained unchanged for some years after the end of the Second World War, until it was repealed in 1953 with the 3rd Criminal Law Reform (*Strafrechtsänderungsgesetz* of August 4, 1953). Medical confidentiality was again included into section 300 of the Penal Code. Several of the German states (*Länder*) in the postwar period demanded, however, as part of their ordinances on residency notification (*Meldeordnungen*), that hospitals immediately report to the police patients who showed patterns of injury that indicated that they had been perpetrators or victims of a crime.[130] In practice, such injuries often stemmed from brawls but also from women's attempts to perform an abortion on themselves. Both the medical and legal professions were highly critical of these regulations, whose roots reached back to the residency laws of the Third Reich, and it seems that hospital doctors widely ignored the requirement to report such cases. Privately practicing doctors were exempt from these regulations anyway. Between 1958 and 1961, the relevant residency ordinances were annulled.[131]

Similarly, the requirement for doctors to report by name any case of abortion and any miscarriage or early birth before the thirty-second week of pregnancy to the health authorities, according to the National Socialist Law for the Prevention of Hereditarily Diseased Offspring, continued to be valid in some of the German *Länder* after World War II. Those federal states that abolished the old regulation after the war issued analogous new ordinances. Bavaria, for example, demanded such reporting until the early 1970s before regulations of this kind were repealed. Since then, doctors and midwives in Germany only have to report stillbirths to the registrar's office for mortality records.[132]

Conclusions

In all three countries—the United States, Britain, and Germany—the question of whether doctors should report cases of illegal abortion to the authorities was a key issue of medico-legal debate. It appears that the majority of medical professionals resisted legal demands for disclosure of such cases, while a minority of doctors, driven by pronatalist sentiments, believed that confidentiality should be breached under these circumstances and did report. The different political and legal contexts had a significant impact on this issue. In many of the American states, the existence of a medical privilege in court provided an argument not to disclose cases of abortion. In Britain, where no such privilege existed, the medical professional organizations were nevertheless protective of confidentiality, fearing that breaches might lead to a loss of doctors' reputation or even expose them to charges of slander or libel. In Germany, before 1933, section 300 of the Reich Penal Code as well as doctors' right to refuse to give evidence according to section 52 of the Code of Criminal Procedure provided strong reasons not to disclose. In all three countries, legal opinion on the question was divided, thus giving doctors additional grounds to preserve secrecy.

With the assumption of power by the National Socialists in Germany, the legal status of medical confidentiality was changed. Removed from the Penal Code and inserted into the Reich Medical Ordinance, which was meant to be an instrument of National Socialist health policy, medical confidentiality was very much weakened in Germany. Moreover, the Law for the Prevention of Hereditarily Diseased Offspring introduced a duty for German doctors to report any abortion to the health authorities. It appears that some doctors tried to subvert this requirement by not stating that a "miscarriage" had been illegally induced. But there were other doctors, alongside legal commentators, who welcomed the new regulations as an authorization to report the "crime" of abortion as an offence against the German *Volk* and the National Socialist state. Even after the end of the Second World War, the German state continued to require doctors to report abortion cases to the health authorities until the relevant regulations were abolished in the 1970s and the reports were reduced to communications to the registrar for statistical purposes.

General Conclusions

As this study has shown, the question of when breaches of medical confidentiality might be justified required an intricate balancing act between individual and collective interests. The issue could not be reduced, however, to a simple dichotomy or a decision between an individual's interest in keeping private information secret and a public interest in disclosure. While such a choice was perceived at the surface of the debates analyzed in this book, medical and legal defenders of confidentiality made a powerful argument that breaches of patients' confidences seriously undermined, perhaps even destroyed, the trust that was essential for an effective doctor–patient relationship. They argued that patients would refrain from consulting a qualified medical practitioner if they felt that knowledge about their conditions and personal circumstances might be disclosed in the interests of others or of the state or be used against them in a court of law. They would therefore seek the help of "quacks" or resort to self-treatment, which would cause harm not only to the individual but to public health as well. Maintaining confidentiality was therefore in the interests of both the collective and the individual. Only rarely was this argument challenged: American law professor John Henry Wigmore was an exception: in 1905, he asked for evidence that the possibility of disclosure in court would deter people from entrusting themselves to a doctor. Yet even he conceded that patients had a real interest in secrecy in cases of venereal disease and abortion.

The notion of privacy, which today is at the core of many debates on medical confidentiality, was still relatively new in the late nineteenth and early twentieth centuries—that is, the period on which this book has focused. Samuel Warren's and Louis Brandeis's now famous paper "The Right to Privacy" appeared in the *Harvard Law Review* in 1890. While it reflected a social climate that was conducive to protecting medical secrecy, it was not commonly referred to even in more specialized medico-legal publications of the time. The key to understanding historical debates on the limits of medical confidentiality lies in the contemporary importance of reputation. Preserving one's reputation was crucial both for patients, who had an interest in their doctors' confidentiality in matters of reproduction and sexual health, and for medical practitioners, who feared for their business if they became known as "informers." Similarly, there were concerns that indiscretion and consequent reputational damage might ruin a clinic's success. Warnings in England during the 1920s that judges, by compelling doctors in divorce proceedings to testify to the venereal disease of one of the spouses, endangered the success of the government's voluntary VD treatment scheme may have been exaggerated. But they reflected a widespread sensitivity on this matter. Attitudes toward abortion conveyed a similar message. The majority of doctors in Britain, the United States, and Germany seem to have been very reluctant to report to the police or public prosecutor if they had become aware of a case of criminal abortion. Pressure to report such cases was resisted out of concern for the woman's reputation and the risk that she might face prosecution, but such pressure was also resisted because doctors needed their female patients' trust in order to sustain their practices.

The pressure to disclose patient details came from two directions: the law and public health. As discussions in all three countries on a medical privilege in court have shown, a considerable body of legal opinions saw doctors' confidentiality as an obstacle to the administration of justice. Efforts to overcome this obstacle manifested themselves in a variety of ways, from exceptions to the statutory medical privilege in criminal proceedings in some American states, to decisions in some German courts that sought to circumvent doctors' entitlement to refuse testimony, to Justice McCardie's ruling in an English court that there were higher considerations than the position of medical men. It appears, however, that the power relations between the legal and medical professions were more balanced in Germany and in those American states that adopted medical privilege statutes than they were in Britain, where the judiciary continued to reject a medical privilege. In Germany, the high political regard in which nineteenth-century medical science was held may have played a role insofar as it also elevated the standing of medical professionals more generally.[1] In Britain, the traditionally high social status of the legal profession probably helped ensure the dominance of lawyers' views in conflicts with medical men. In regard to interactions between the medical and legal professions in Britain from the late eighteenth to the twentieth century, Imogen Goold and Catherine Kelly have observed an "inability of medical and legal intellectual structures to communicate with each other." Resulting conflicts, they note, were in most cases "won by the legal system leading to the conclusion that medicine's increased utility to that system caused its own subjugation."[2] In the United States, regular medical practitioners struggled for a long time for legal recognition of their profession and formal demarcation from homeopaths, herbalists, and other alternative practitioners through new restrictive licensing legislation after earlier licensing laws had been repealed in the 1830s and 1840s.[3] Modern medical licensing began to be introduced in the United States only in the early 1880s, starting with the Board of Health Act of West Virginia of 1881–82.[4] The sociopolitical influence of the American medical profession seems to have varied from state to state. More general differences in the balance of medico-legal relationships are reflected in the variety of American regulations on doctors' testimony in court.

Some legal critics, as well as some pronatalist medical campaigners, held that doctors maintaining confidentiality about abortions were aiders and abettors of a crime. Undoubtedly, abortion posed a serious moral problem for doctors as long as the practice was criminalized—which was the case for all three countries during the late nineteenth and the first half of the twentieth century (and beyond). Most doctors, it appears, decided not to report. Even in National Socialist Germany, where the reporting of any miscarriage or abortion to the health authorities was made mandatory, there were attempts to avoid incriminating the women concerned. Usually, only if a woman had died from an abortion procedure were doctors ready to report the case to the police so that action might be taken against a dangerous abortionist.

The public health argument for disclosure was particularly prominent in the debates on venereal disease notification and the warning of contacts. Many critics, including prominent medical commentators such as Albert Neisser, reminded doctors of their duty not only to the individual VD patient under their care but also to the patient's social environment and to the health of the public in general. Those who argued for the priority of public health pointed to the existing notification laws for other contagious diseases. Such laws showed how public interest in protection against infection superseded the individual's interest in not being stigmatized as a danger to his or her environment and in avoiding the hardships of isolation. The approaches taken to solve the VD problem differed among the three countries. While by 1919 the United States had adopted a policy of notification of cases of venereal disease to the health authorities, at least in anonymized form, Britain followed a strategy of voluntary confidential treatment. In 1927, Germany enacted a law that constituted a middle way: only VD patients who failed to comply with medical treatment and regular controls had to be reported, as well as those whose occupation and personal circumstances (e.g., prostitution) created a high risk of infecting others.

As this historical study has demonstrated, however, there were no easy solutions to the conflicts between, on the one side, the traditional value of medical confidentiality and the new notion of a right to privacy and, on the other, the increasing demands for medical information and expertise in legal systems and in protecting public health. Indeed, many of the historical dilemmas analyzed in this book are in essence still with us today. What has changed is rather the emphasis on certain types of issues connected with confidentiality. While the problems of the nineteenth and early twentieth centuries were mostly concerned with the individual medical practitioner, who had to decide between secrecy and disclosure, today's prominent issues are related to the secure management and sharing of electronic health data, including genetic information. Again, the potential of medical information to improve public health has to be balanced against the risk of harming people by inadvertently or deliberately revealing sensitive aspects of their personal, private lives.

Recent events in England have shown how sensitive this issue is. In February 2014, NHS (National Health Service) England and the Health and Social Care Information Centre (HSCIC, a non-departmental public body established in 2012 that collects, holds, and selectively releases information from health and social care institutions) temporarily put their "care.data" project on hold amid privacy concerns and the perception that the general public had been insufficiently informed. The plan was to extract health data from the IT systems of all NHS GP medical practices in England, an enormous amount of data, including patients' NHS numbers (though not their names), dates of birth, postcodes, diagnoses, prescriptions, and procedures such as vaccinations. Patients were supposed to be given the opportunity to opt out from the scheme, and their data were to be used for auditing and commissioning of health care provision, research, and public health purposes only. Leaflets were delivered to households to inform them about the project, but many people subsequently claimed never to have received this information.[5]

On February 23, 2014, *The Telegraph* reported, on the basis of Official Information Act requests, that the precursor of the HSCIC, the NHS Information Centre, had sold the data of forty-seven million NHS hospital inpatients from 1997 to 2010 to insurance companies.[6] In the wake of this revelation, the ethical issues and safeguards for the collection and sharing of health data were scrutinized and intensely discussed, involving, among others, the House of Commons Health Committee, the All-Party Parliamentary Group (APPG) for Patient and Public Involvement in Health and Social Care, and the Nuffield Council on Bioethics (an important think tank for policies in health and biomedical science). A key conclusion of the APPG report, published in November 2014, was that patients must be properly informed about their rights and choices regarding the "care.data" scheme, as well as about its potential benefits, so that they can consent within an opt-out system to their data being collected.[7] A legislative amendment in the Care Act of 2014 required the formation of the Confidentiality Advisory Group to the HSCIC, tasked to advise the center on disclosures.[8] The Nuffield Council report, which came out early in 2015, made several recommendations, including that an independent group of participant representatives should convene to develop a public statement on how data held by the HSCIC should be used.[9] While the HSCIC could proceed to pilot the care.data project in a number of "pathfinder practices" in selected areas of England,[10] the thrust of the proposed solution was to seek the broad consent of the public as the "data subjects." In other words, informed consent to disclose personal health information was (and is) the approach taken to solve the tensions between demands for privacy and the promises of improved health care and other socioeconomic advantages.

The seeking of consent to disclose patient information is, as this book has illustrated, *as such* not new. Patient waivers of confidentiality were recognized in American courts as a means to enable doctors to testify in spite of the medical privilege, and VD patients in Scotland were asked to inform past sexual contacts by giving them medical letters for their doctors. However, such practices were part of a different framework, where doctors were the relevant custodians of confidential patient information and where patients were rarely perceived as autonomous decision makers in matters concerning their health. What we can learn from the historical record is that there may be good reasons to disclose patient information under specific circumstances *without consent*—for example, to warn close contacts of a danger of infection with a serious contagious disease if the patient is uncooperative and refuses to inform those contacts. There may also be a good reason for doctors to testify without a patient's consent in criminal trials if the medical evidence may avert a miscarriage of justice. On the other hand, as the historical resistance of doctors to reporting abortions, even in a time when abortion was illegal, shows, preserving trust in the physician–patient relationship can be a powerful reason not to disclose. Moreover, when public health arguments seem to mandate the reporting of cases, one needs to ask whether anonymized information may be sufficient and how that anonymity can best be protected. While the history of medical confidentiality cannot provide specific normative answers to present problems of patient privacy, it can make us more sensitive to the fundamental ethical issues at stake. I hope that this book will make a contribution toward this aim.

In looking back on the history of confidentiality and patient privacy, we should also be aware, however, of the changing ethical and legal landscape. As Jean McHale, Birmingham professor of health care law, has recently suggested with regard to the care. data debate, we may be at a critical turning point where we should "move from a focus on unauthorised disclosure to return to first principles and reframing this area [of privacy and confidentiality of health care information] in terms of patients' rights and patient autonomy."[11] This could mean that, in the future, patients might control and manage their personal health information online and decide on a case-by-case basis for which public health purposes or medical research projects their data can be released and used. If such policies were to be implemented, we might see a scenario where the ethical conflict between secrecy and disclosure in the public interest, which has been the focus of this book, will be significantly shifted from the health professional to the autonomous patient. A future history of the topic would then need to assess whether and how patients have been empowered to take on this role as guardians of their health information and how they have dealt with this new responsibility.

Acknowledgments

I gratefully acknowledge the support of the Leverhulme Trust by awarding me the research fellowship (RF-2013-005) that enabled me to write this book. My editor at the University of Chicago Press (UCP), Karen Merikangas Darling, gave useful and critical advice throughout the process. Thanks likewise go to the anonymous reviewers for UCP and to my colleague Lutz Sauerteig at Durham's Centre for the History of Medicine and Disease for their helpful suggestions. Sebastian Pranghofer provided efficient research assistance during my earlier work on medical confidentiality as part of the center's Wellcome Trust Enhancement and Strategic Awards. Thanks also to Arlene Shaner for her friendly advice during my visit to the Library of the New York Academy of Medicine. Preliminary research findings have been presented at conferences, research seminars, and public lectures in Durham, Glasgow, Halifax (Nova Scotia), and York. I am grateful for the feedback obtained at these occasions. Furthermore, I thank Durham University for granting me research leave during Easter Term 2015 to carry out final revisions of the manuscript, Evan White of UCP for copyright advice, and Uitgeverij Paris for permission to reuse my article "Preserving Confidentiality or Obstructing Justice? Historical Perspectives on a Medical Privilege in Court" (*Journal of Medical Law and Ethics*, 3 [2015]) as a basis for the first chapter of this book. Last but not least,

I would like to thank my wife, Jillian, for commenting on my text from her perspective as a barrister, and my daughter, Sophie, now an undergraduate student in law, for helping with journal analyses during "work experience" while still at school.

Notes

Introduction

1. See, for example, British Medical Association, *Medical Ethics Today: The BMA's Handbook of Ethics and Law*, 2nd ed. (London: BMJ Books, 2004), 165–66; Shaun Pattinson and Deryck Beyleveld, "Confidentiality and Data Protection," in *Principles of Medical Law*, 3rd ed., ed. Andrew Grubb, Judith Laing, and Jean McHale (Oxford: Oxford University Press, 2010), 651–52.

2. "What I may see or hear in the course of the treatment or even outside of the treatment in regard to the life of men, which on no account one must spread abroad, I will keep to myself holding such things shameful to be spoken about" (trans. Ludwig Edelstein).

3. Decision in the case of Dr. J. Marion Sims, June 15, 1870, Scrapbook for Committee on Ethics, Archive of the New York Academy of Medicine, New York.

4. Letter of Dr. Sims to the Editor of the *New York Times*, October 3, 1869, ibid.

5. Emma Stebbins, *Charlotte Cushman: Her Letters and Memories of Her Life* (Boston: Houghton, Osgood & Company, 1879), 230.

6. Dr. Richardson to Dr. Finnell, New York Academy of Medicine, Charges against J. Marion Sims, undated, Scrapbook for Committee on Ethics, Archive of the New York Academy of Medicine, New York. Cf. Isaac Hays, "Code of Ethics" (1847), in *The Codification of Medical Morality*, vol. 2: *Anglo-American Medical Ethics and Medical Jurisprudence in the Nineteenth Century*, ed. Robert Baker (Dordrecht: Kluwer Academic, 1995), 75–76, 80.

7. Letter of J. Marion Sims to the Committee on Ethics of the New York Academy of Medicine, April 4, 1870, Scrapbook for Committee on Ethics, Archive of the New York Academy of Medicine, New York.

8. Ibid.

9. The case has also been briefly discussed by Philip Van Ingen, *The New York Academy of Medicine: Its First Hundred Years* (New York: Columbia University Press, 1949), 144–45, and by Robert Baker, *Before Bioethics: A History of American Medical Ethics from the Colonial Period to the Bioethics Revolution* (New York: Oxford University Press, 2013), 108–9.

10. Ludwig Ebermayer, "Die Unruhen in Berlin-Moabit und das Zeugnisverweigerungsrecht der Aerzte," *Ärztliches Vereinsblatt für Deutschland* 37 (1910): 828–29. On Ebermayer, who became the highest ranking public prosecutor in Germany (*Oberreichsanwalt*) in 1921, see Andreas Michael Staufer, *Ludwig Ebermayer: Leben und Werk des höchsten Anklägers in der Weimarer Republik unter besonderer Berücksichtigung seiner Tätigkeit im Medizin- und Strafrecht* (Leipzig: Leipziger Universitätsverlag, 2010).

11. Albert Hellwig, "Die Beschlagnahme ärztlicher Krankenjournale nach geltendem und künftigem Recht," *Deutsche Medizinische Wochenschrift* 36 (1910): 2152–53.

12. Jutta Steinberg-Copek, "Berufsgeheimnis und Aufzeichnungen des Arztes im Strafverfahren" (Jur. Diss., Freie Universität Berlin, 1968), 17, 143–44, 154.

13. Rolf Lamprecht, "Wieviel ist das Arztgeheimnis noch wert? Zur Güterabwägung zwischen Privatsphäre und Strafrechtspflege, erläutert am Memminger Exempel," *Zeitschrift für Rechtspolitik* 22, no. 8 (1989): 290–93; Sabine Michalowski, *Medical Confidentiality and Crime* (Aldershot: Ashgate, 2003), 122–23; Bundesverfassungsgericht—Entscheidungen, 2 BvR 291/92 (decision of May 22, 2000), http://www.bundesverfassungsgericht.de/entscheidungen/rk20000522_2bvr029192.html.

14. See Michalowski, *Medical Confidentiality and Crime.*

15. Brief discussions of these issues for imperial Germany can be found in Siegfried Placzek, *Das Berufsgeheimnis des Arztes*, 3rd enlarged and revised edition (Leipzig: Verlag von Georg Thieme, 1909), 175–83, 211–15; and Andreas-Holger Maehle, *Doctors, Honour and the Law: Medical Ethics in Imperial Germany* (Basingstoke: Palgrave Macmillan, 2009), 62–64. For a modern survey and discussion of confidentiality issues in mental health care in the United States, see Ralph Slovenko, *Psychotherapy and Confidentiality. Testimonial Privileged Communication, Breach of Confidentiality, and Reporting Duties* (Springfield: Charles C. Thomas, 1998).

16. Raymond Villey, *Histoire du Secret Médical* (Paris: Seghers, 1986).

17. Amy L. Fairchild, Ronald Bayer, and James Colgrove, with Daniel Wolfe, *Searching Eyes: Privacy, the State, and Disease Surveillance in America* (Berkeley: University of California Press, 2007).

18. Angus H. Ferguson, *Should a Doctor Tell? The Evolution of Medical Confidentiality in Britain* (Farnham: Ashgate, 2013).

19. For example, Angus McLaren, "Privileged Communications: Medical

Confidentiality in Late Victorian Britain," *Medical History* 37 (1993): 129–47; Andrew A. G. Morrice, "'Should the Doctor Tell?' Medical Secrecy in Early Twentieth-Century Britain," in *Medicine, Health and the Public Sphere in Britain, 1600–2000*, ed. Steve Sturdy (London: Routledge, 2002), 60–82; Andreas-Holger Maehle, "Protecting Patient Privacy or Serving Public Interests? Challenges to Medical Confidentiality in Imperial Germany," *Social History of Medicine* 16 (2003): 383–401; Angus H. Ferguson, "The Lasting Legacy of a Bigamous Duchess: The Benchmark Precedent for Medical Confidentiality," *Social History of Medicine* 19 (2006): 37–53; Ferguson, "Exploring the Myth of a Scottish Privilege: A Comparison of the Early Development of the Law on Medical Confidentiality in Scotland and England," in *Medicine, Law and Public Policy in Scotland c. 1850–1990. Essays Presented to Anne Crowther*, ed. Mark Freeman, Eleanor Gordon, and Krista Maglen (Dundee: Dundee University Press, 2011), 125–40; Ferguson, "Medical Confidentiality in the Military," in *Military Medical Ethics for the 21st Century*, ed. Michael L. Gross and Don Carrick (Farnham: Ashgate, 2013), 209–24; Ferguson, "The Role of History in Debates Regarding the Boundaries of Medical Confidentiality and Privacy," *Journal of Medical Law and Ethics* 3 (2015): 65–81; Frances L. Bernstein, "Behind the Closed Door: VD and Medical Secrecy in Early Soviet Medicine," in *Soviet Medicine: Culture, Practice, and Science*, ed. F. L. Bernstein, Christopher Burton, and Dan Healy (DeKalb: Northern Illinois University Press, 2010), 92–110. Professional confidentiality in Soviet medicine has also been recently discussed in Igor J. Polianski, *Das Schweigen der Ärzte: Eine Kulturgeschichte der sowjetischen Medizin und ihrer Ethik* (Stuttgart: Franz Steiner Verlag, 2015). Furthermore, see my preliminary binational comparison: Andreas-Holger Maehle and Sebastian Pranghofer, "Medical Confidentiality in the Late Nineteenth and Early Twentieth Centuries: An Anglo-German Comparison," *Medizinhistorisches Journal* 45 (2010): 189–221.

20. See, for example, Pattinson and Beyleveld, "Confidentiality and Data Protection"; Franziska Lang, *Das Recht auf informationelle Selbstbestimmung des Patienten und die ärztliche Schweigepflicht in der gesetzlichen Krankenversicherung* (Baden-Baden: Nomos Verlagsgesellschaft, 1997); Rosemary Jay, *Data Protection Law and Practice*, 4th ed., 1st suppl. (Andover: Sweet & Maxwell, Thomson Reuters, 2014); Jean McHale, "From *X v Y* to care.data and Beyond: Health Care Confidentiality and Privacy in the C21st: A Critical Turning Point?," *Journal of Medical Law and Ethics* 3 (2015): 109–33; Michael Soljak, "Big Voice or Big Data? The Difficult Birth of care.data," *Journal of Medical Law and Ethics* 3 (2015): 135–42.

Chapter One

1. For interpretations of the precept of secrecy in the Hippocratic Oath, see Steven H. Miles, *The Hippocratic Oath and the Ethics of Medicine* (Oxford: Oxford University Press, 2004), 149–57.

2. This chapter expands and develops my preliminary article, Andreas-Holger Maehle, "Preserving Confidentiality or Obstructing Justice? Historical Perspectives on a Medical Privilege in Court," *Journal of Medical Law and Ethics* 3 (2015): 91–108.

3. For detailed discussions of this case, see Angus H. Ferguson, "The Lasting Legacy of a Bigamous Duchess: The Benchmark Precedent for Medical Confidentiality," *Social History of Medicine* 19 (2006): 37–53; Danuta Mendelson, "The Duchess of Kingston's Case, the Ruling of Lord Mansfield and Duty of Medical Confidentiality in Court," *International Journal of Law and Psychiatry* 35 (2012): 480–89; and Angus H. Ferguson, *Should a Doctor Tell? The Evolution of Medical Confidentiality in Britain* (Farnham: Ashgate, 2013), 13–28. See also the account of the trial in Claire Gervat, *Elizabeth: The Scandalous Life of the Duchess of Kingston* (London: Century, 2003), 144–54.

4. *The Trial of Elizabeth Duchess Dowager of Kingston for Bigamy, Before the Right Honourable The House of Peers, in Westminster-Hall, in Full Parliament . . . Published by Order of the House of Peers* (London: Charles Bathurst, 1776), 119; see also Thomas Bayly Howell, *A Complete Collection of State Trials*, vol. 20 (London: Longman, Hurst, Rees etc., 1816), 572.

5. *The Trial of Elizabeth*, 120; Howell, *State Trials*, 573.

6. Ferguson, "Lasting Legacy," 45–46; Gervat, *Elizabeth*, 153–55.

7. Buller cited in Howell, *State Trials*, 575–76.

8. John Henry Wigmore, *A Treatise on the System of Evidence in Trials at Common Law*, vol. 4 (Boston: Little, Brown, and Company, 1905), 3347–48; W. A. Purrington, "An Abused Privilege," *Columbia Law Review* 6 (1906): 390–92; Ferguson, "Lasting Legacy," 46–47. Mendelson has argued that Lord Mansfield's ruling against a medical privilege in court was misunderstood, as it was meant to refer only to matters of public knowledge, such as the existence of a marriage and the birth of a child, not to information obtained within the nonpublic relationship between patient and medical practitioner. See Mendelson, "The Duchess of Kingston's Case," 484.

9. Ferguson, "Lasting Legacy," 47–48.

10. Ferguson, *Should a Doctor Tell?*, 25; Angus H. Ferguson, "Exploring the Myth of a Scottish Privilege: A Comparison of the Early Development of the Law on Medical Confidentiality in Scotland and England," in *Medicine, Law and Public Policy in Scotland c. 1850–1990. Essays Presented to Anne Crowther*, ed. Mark Freeman, Eleanor Gordon, and Krista Maglen (Dundee: Dundee University Press, 2011), 130.

11. Ferguson, "Lasting Legacy," 42; Ferguson, *Should a Doctor Tell?*, 21.

12. John Gregory, *Lectures on the Duties and Qualifications of a Physician*, new edition, corrected and enlarged (London: W. Strahan and T. Cadell, 1772), 26. For a detailed discussion of Gregory's *Lectures* and their context, see Laurence B.

McCullough, *John Gregory and the Invention of Professional Medical Ethics and the Profession of Medicine* (Dordrecht: Kluwer Academic, 1998).

13. Thomas Percival, "Medical Ethics; or a Code of Institutes and Precepts, Adapted to the Professional Conduct of Physicians and Surgeons" (1803), in *Percival's Medical Ethics*, ed. Chauncey D. Leake (Huntington, NY: Robert E. Krieger, 1975), 90. For an interpretation of Percival's medical ethics, see Robert Baker, "Deciphering Percival's Code," in *The Codification of Medical Morality*, vol. 1: *Medical Ethics and Etiquette in the Eighteenth Century*, ed. Robert Baker, Dorothy Porter, and Roy Porter (Dordrecht: Kluwer Academic, 1993), 179–211.

14. Percival, "Medical Ethics," 63.

15. Ibid., 161–65. Emphasis in the original.

16. The prisoner was acquitted, however, on other grounds; cf. Theodric Romeyn Beck and John B. Beck, *Elements of Medical Jurisprudence*, 12th ed., revised by C. R. Gilman, vol. 2 (Philadelphia: J. B. Lippincott and Co., 1863), 965; Purrington, "Abused Privilege," 391–92; Ferguson, *Should a Doctor Tell?*, 24.

17. Critical views regarding the lack of a medical privilege were expressed, besides Buller, by Lord Chancellor Brougham in *Greenough v. Gaskell* (1833) and by the Scottish surgeon and professor of medical jurisprudence John Gordon Smith (1825); see Ferguson, *Should a Doctor Tell?*, 27; and Mendelson, "The Duchess of Kingston's Case," 485, 487.

18. N.Y. Rev. Stats. 1st ed. Vol. 2, p. 406, Pt. III, Ch. VII, Tit. III, Art. 8, Sec. 73. Cited in Purrington, "Abused Privilege," 392. See also Ralph Slovenko, *Psychotherapy and Confidentiality. Testimonial Privileged Communication, Breach of Confidentiality, and Reporting Duties* (Springfield: Charles C. Thomas, 1998), 22.

19. Clinton DeWitt, *Privileged Communications between Physician and Patient* (Springfield: Charles C. Thomas, 1958), 15.

20. Commissioners on Revision of the Statutes of New York III, 737 (1836). Cited in Wigmore, *System of Evidence*, 3349–50.

21. Robert Baker, *Before Bioethics: A History of American Medical Ethics from the Colonial Period to the Bioethics Revolution* (New York: Oxford University Press, 2013), 118.

22. Wigmore, *System of Evidence*, 3348.

23. G. W. Field and John B. Uhle, "Privileged Communications," *The American Law Register* 37 (1889): 1–21.

24. Arkansas, California, Colorado, Idaho, Indiana, Iowa, Kansas, Michigan, Minnesota, Missouri, Montana, Nebraska, Nevada, New York, North Carolina, North Dakota, Ohio, Oklahoma, Oregon, Pennsylvania, South Dakota, Utah, Washington, Wisconsin, and Wyoming.

25. Tracy C. Becker, "Observations Concerning the Law of Privileged Communications between Physician and Patient, as Applicable to the Duties of Railway Surgeons," *Journal of the American Medical Association (JAMA)* 26 (1896):

1065–66. At the time *no* restrictions on the disclosures that a physician could be compelled to make in court existed in Alabama, Arizona, Connecticut, Delaware, District of Columbia, Florida, Georgia, Illinois, Kentucky, Louisiana, Maine, Maryland, Massachusetts, Mississippi, New Hampshire, New Jersey, New Mexico, Rhode Island, South Carolina, Tennessee, Texas, Vermont, Virginia, and West Virginia. Cf. Charles A. Boston, "The Law of Evidence concerning Confidential Communications between Physician and Patient," in *Medical Jurisprudence, Forensic Medicine and Toxicology*, ed. R. A. Witthaus and Tracy C. Becker, vol. 1 (New York: William Wood & Company, 1894), 94.

26. "Privileged Communications," *JAMA* 62 (1914): 1350–51.

27. Field and Uhle, "Privileged Communications," 11–12; "Medical Confidences," *JAMA* 33 (1899): 1431.

28. California, Idaho, Minnesota, Montana, North Dakota, Oregon, South Dakota, Utah, and Washington. Cf. Becker, "Observations," 1066; Boston, "Law of Evidence," 96; Samuel I. Jacobs, "Evidence: Privileged Communications between Physician and Patient in California: Cal. Code Civ. Proc. § 1881 (4)," *California Law Review* 20 (1932): 304.

29. "Physicians and Privileged Communications," *JAMA* 51 (1908): 1170.

30. Medical malpractice suits became frequent in America from the 1840s; see James C. Mohr, "American Medical Malpractice Litigation in Historical Perspective," *JAMA* 283 (2000): 1731–37.

31. Hugh Emmett Culbertson, *Medical Men and the Law. A Modern Treatise on the Legal Rights, Duties and Liabilities of Physicians and Surgeons* (Philadelphia: Lea & Febiger, 1913), 299–300; "Privilege Waived by Bringing of Action for Malpractice—Evidence of Defense by Medical Society Not Admissible," *JAMA* 58 (1912): 511; "Construction of Statute Relative to Privileged Communications—What Adjudged Waivers—Application to Several Physicians," *JAMA* 61 (1913): 1837–38; "Privileged Communications under Amended Statute," *JAMA* 64 (1915): 1446; "Waiver of Privilege by Plaintiffs," *JAMA* 79 (1922): 2250; "Waiver of Privilege in Personal Injury Case," *JAMA* 80 (1923): 1172; "When Doctrine of 'Res Ipsa Loquitur' Applies," *JAMA* 91 (1928): 1919–20; Lloyd Paul Stryker, *Courts and Doctors* (New York: Macmillan, 1932), 34–40. See, however, the personal injury case of *Smart v. Kansas City*, where the Supreme Court of Missouri upheld the patient's decision to exclude the doctor's testimony by refusing to waive medical secrecy: "Privilege Not Waived by Bringing Action for Personal Injuries," *JAMA* 50 (1908): 723–24. See also the malpractice trial of *Hartley v. Calbreath*, where another physician offered as a witness by the defendant was, on the plaintiff's objection, not permitted by the court to testify, because that physician's communications with the patient (plaintiff) were privileged under the Missouri statute. Cf. "Extent of Waiver of Privilege by Bringing Action for Malpractice," *JAMA* 50 (1908): 1217–18.

32. "Privilege Attaches to Hospital Records—When Treated as Public Records," *JAMA* 79 (1922): 2188.

33. *People v. De France* (decided April 2, 1895). Cf. "Communications Made to Dentists Are Not Privileged," *JAMA* 24 (1895): 949–50.

34. "Doctrine of Privileged Communications," *JAMA* 54 (1910): 2094.

35. "Rule as to Privileged Communications Applied to Roentgenologist," *JAMA* 71 (1918): 2019.

36. Specifically, in 1891, 1892, 1899, and 1904. See Purrington, "Abused Privilege," 402–3; "Privileged Communications," *JAMA* 62 (1914): 1351.

37. "Physical Examination and Privilege Waiving," *JAMA* 48 (1907): 903.

38. Boston, "Law of Evidence," 98, 107–8; Stryker, *Courts and Doctors*, 38–39. In contrast to this, the Supreme Court of Wisconsin took the view that only the patient can waive the privilege of secrecy and that therefore after the patient's death "the physician's lips are forever sealed under all circumstances." Cf. "Wisconsin Doctrine as to Privileged Communications," *JAMA* 47 (1904): 1577. In the same vein, the Supreme Court of Mississippi ruled in a case of contested will that it was error to admit the testimony of a physician about the deceased patient's senile dementia. Cf. "Statutes Relative to Privileged Communications and Vital Statistics," *JAMA* 79 (1922): 325.

39. Purrington, "Abused Privilege," 403.

40. See Bettina Wahrig and Werner Sohn (eds.), *Zwischen Aufklärung, Policey und Verwaltung: Zur Genese des Medizinalwesens 1750–1850* (Wiesbaden: Harrassowitz Verlag, 2003).

41. Bjarne Exner, "Das Berufsgeheimnis des Arztes gemäß § 300 des Str. G. B." (Jur. Diss., Ruprecht-Karls-Universität Heidelberg, 1909), 11.

42. Preußisches Allgemeines Landrecht, Teil 2, Tit. 20, § 505. Cf. Siegfried Placzek, *Das Berufsgeheimnis des Arztes*, 3rd enlarged and revised edition (Leipzig: Verlag von Georg Thieme, 1909), 2.

43. Claudia Huerkamp, *Der Aufstieg der Ärzte im 19. Jahrhundert. Vom gelehrten Stand zum professionellen Experten: Das Beispiel Preußens* (Göttingen: Vandenhoeck & Ruprecht, 1985), 27.

44. Placzek, *Berufsgeheimnis* (1909), 49. The French legal requirement of professional secrecy was taken to be "absolute," applying also to evidence in criminal trials. See, for example, "Professional Secrecy," *JAMA* 35 (1900): 104.

45. Placzek, *Berufsgeheimnis* (1909), 7.

46. Ibid., 4–5.

47. Reichs-Strafgesetzbuch (RStGB), § 300. Cf. Placzek, *Berufsgeheimnis* (1909), 2.

48. Huerkamp, *Der Aufstieg der Ärzte*, 210.

49. Placzek, *Berufsgeheimnis* (1909), 29.

50. Gesetz, betr. die Bekämpfung gemeingefährlicher Krankheiten, June 30,

1900. Cf. Placzek, *Berufsgeheimnis* (1909), 30. See also chapter 2, under "A Landmark Decision in Germany, Its Context, and Its Consequences."

51. Exner, "Das Berufsgeheimnis," 41; Hans Weizmann, "Das Berufsgeheimnis. (§ 300 RStGB.)" (Jur. Diss., Universität Breslau, 1909), 25.

52. Heinrich Schmidt, *Das ärztliche Berufsgeheimnis* (Jur. Diss., Universität Leipzig, Jena: Gustav Fischer, 1907), 19–20.

53. J. Liebmann, *Die Pflicht des Arztes zur Bewahrung anvertrauter Geheimnisse* (Frankfurt a. M.: Joseph Baer and Co., 1886), 6–8; Placzek, *Berufsgeheimnis* (1909), 32.

54. Hermann Wichmann, *Das Berufsgeheimnis als Grenze des Zeugenbeweises. Ein Beitrag zur Lehre von den Beweisverboten* (Jur. Diss., Universität Göttingen, 2000; Frankfurt/Main: Peter Lang, 2000), 205–6.

55. Important for this view was a decision of the District Court of Hamburg on June 24, 1899; see W. Mittermaier, "Gutachten über § 300 R.St.G.B.," *Zeitschrift für die gesamte Strafrechtswissenschaft* 21 (1902): 229–30. See also Friedrich Ottomar Jummel, "Der § 300 Str.G.B., ein Versuch seiner Auslegung" (Jur. Diss., Universität Leipzig, 1903), 54.

56. H.-J. Rieger, "Zur geschichtlichen Entwicklung der ärztlichen Schweigepflicht," *Deutsche Medizinische Wochenschrift* 100 (1975): 1867–68.

57. Cited in Exner, "Das Berufsgeheimnis," 16. See also Fritz Sauter, *Das Berufsgeheimnis und sein strafrechtlicher Schutz. (§ 300 R.St.G.B.)* (Breslau: Schletter'sche Buchhandlung, 1910), 32–33.

58. *Entscheidungen des Reichsgerichts in Strafsachen* 13 (1886): 60–64 (decision of October 22, 1885).

59. William Mawdesley Best, *The Principles of the Law of Evidence with Elementary Rules for Conducting the Examination and Cross-Examination of Witnesses*, first American, from the sixth London edition of John A. Russell, by James Appleton Morgan, vol. 2 (New York: Cockcroft & Company, 1878), 987–88. Best's handbook saw twelve editions. In addition to Britain and the United States, it was also published in India and Germany.

60. Boston, "Law of Evidence," 134.

61. Ibid.

62. *Nelson v. Village of Oneida*, 156 N.Y. 219. Cited in W. A. Purrington, *A Review of Recent Legal Decisions Affecting Physicians, Dentists, Druggists and the Public Health* (New York: E. B. Treat & Company, 1899), 47.

63. Purrington, "Abused Privilege," 393–94, 422. See also William A. Purrington, "Of Certain Legal Relations of Physicians and Surgeons to Their Patients and to One Another," in *A System of Legal Medicine* (1894), ed. Allan McLane Hamilton and Lawrence Godkin, 2nd ed., vol. 1 (New York: E. B. Treat & Company, 1900), 626.

64. Becker, "Observations," 1067.

65. Wigmore, *System of Evidence*, 3350. See also Daniel W. Shuman, "The Origins of the Physician-Patient Privilege and Professional Secret," *Southwestern Law Journal* 39 (1985/86): 663.

66. Wigmore, *System of Evidence*, 3350.

67. Ibid., 3351.

68. Ibid., 3352. Similar concerns were also highlighted by William C. Tait, "Professional Secrecy and Its Legal Aspects," *JAMA* 33 (1899): 460, and by C. F. Taeusch, "Should the Doctor Testify?," *International Journal of Ethics* 38 (1928): 410–12.

69. Cf. Boston, "Law of Evidence," 96, 107; "Privilege Is That of Patient and Not of Physician," *JAMA* 73 (1919): 1083; Jacobs, "Evidence," 307–8.

70. *Treanor v. Man. Ry. Co.* (Pryor, J., 1891) 16 N.Y. Sup. 536. Cited in Purrington, "Abused Privilege," 404. See also Purrington, "Of Certain Legal Relations," 628–29, and Boston, "Law of Evidence," 114.

71. Purrington, "Abused Privilege," 405.

72. "Waivers of Privilege," *JAMA* 57 (1911): 413.

73. "Construction of Statute as to Privileged Communications and Waivers of Privilege, Especially as between Several Physicians," *JAMA* 57 (1911): 1158.

74. "Privilege Not Waived nor Statute Modified," *JAMA* 60 (1913): 1830.

75. Daniel R. Brower, "The Necessity of Granting Privileged Communications to the Medical Profession in the State of Illinois," *JAMA* 27 (1896): 1271–73.

76. Isaac Hays, "Code of Ethics" (1847), in *The Codification of Medical Morality*, vol. 2: *Anglo-American Medical Ethics and Medical Jurisprudence in the Nineteenth Century*, ed. Robert Baker (Dordrecht: Kluwer Academic, 1995), 75.

77. Percival's language was used throughout the Code of Ethics; see Robert Baker, "The Historical Context of the American Medical Association's 1847 *Code of Ethics*," in *The Codification of Medical Morality*, vol. 2: *Anglo-American Medical Ethics and Medical Jurisprudence in the Nineteenth Century*, ed. Robert Baker (Dordrecht: Kluwer Academic, 1995), 53.

78. Hays, "Code of Ethics," 75–76.

79. Henry D. Holton, Leartus Connor, Daniel T. Nelson, and Benjamin Lee, "Code of Medical Ethics and Etiquette of the American Medical Association. Report of Majority Committee," *JAMA* 22 (1894): 507.

80. Baker, *Before Bioethics*, 214.

81. Ibid., 215–18.

82. American Medical Association, "Proceedings of the Fifty-Fourth Annual Session, Held at New Orleans, May 5, 6, 7 and 8, 1903," *JAMA* 40 (1903): 1379.

83. The context of Flint's publication in 1882 was the intraprofessional debate on the AMA code's "consultation clause," which prohibited consultation with homeopaths and other unorthodox practitioners, and the wider question whether

a code of medical ethics was necessary at all. See John Harley Warner, "The 1880s Rebellion against the AMA Code of Ethics: 'Scientific Democracy' and the Dissolution of Orthodoxy," in *The American Medical Ethics Revolution. How the AMA's Code of Ethics Has Transformed Physicians' Relationships to Patients, Professionals, and Society*, ed. Robert B. Baker, Arthur L. Caplan, Linda L. Emanuel, and Stephen R. Latham (Baltimore: Johns Hopkins University Press, 1999), 52–69; Owen Whooley, *Knowledge in the Time of Cholera: The Struggle over American Medicine in the Nineteenth Century* (Chicago: University of Chicago Press, 2013), 96–105; Baker, *Before Bioethics*, 153–55, 205–11.

84. Austin Flint, *Medical Ethics and Etiquette. The Code of Ethics Adopted by the American Medical Association, with Commentaries* (New York: D. Appleton and Company, 1893), 15–17.

85. Flint, *Medical Ethics*, 20. See also Baker, *Before Bioethics*, 209–10.

86. Samuel C. Busey, "The Code of Ethics," *JAMA* 34 (1900): 256.

87. "Professional Communications to Physicians Should Be Privileged," *JAMA* 28 (1897): 374.

88. John B. Hamilton, "'Medical' Legislation and How to Obtain It," *JAMA* 28 (1897): 1005–6.

89. James B. Baird, "The Medical Witness. His Rights and Wrongs in Courts of Justice," *JAMA* 32 (1899): 1149. Other pleas for legislation to protect medical confidentiality in court were, regarding the District of Columbia, "Medical Confidences and Professional Honor," *JAMA* 26 (1896): 783–84; with regard to New Jersey, Daniel Strock, "A Plea for the Physician on the Witness Stand," *Transactions of the Medical Society of New Jersey* 1901: 169–77; and, generally, "Medical Confidences and Medical Testimony," *JAMA* 31 (1898): 309–10.

90. Samuel D. Warren and Louis D. Brandeis, "The Right to Privacy," *Harvard Law Review* 4 (1890): 193–220.

91. Whether the cases cited by Warren and Brandeis were actually precedents showing a right to privacy is questionable, because the decisions in these cases were largely based on property rights. See Walter F. Pratt, *Privacy in Britain* (Lewisburg: Bucknell University Press, 1979), 19–37.

92. Pratt, *Privacy in Britain*, 19; Raymond Wacks, *Privacy and Media Freedom* (Oxford: Oxford University Press, 2013), 53.

93. Warren and Brandeis, "Right to Privacy," 196.

94. They briefly referred, though, to a hypothetical example of Lord Eldon in 1820, that "if one of the late king's [George III] physicians had kept a diary of what he heard and saw, the court would not, in the king's lifetime, have permitted him to print and publish it." Warren and Brandeis, "Right to Privacy," 205.

95. An exception was William C. Tait, LL.B., PhD, of San Francisco, who advised doctors to observe strict secrecy in the Hippocratic tradition and to give evidence only with the consent of the patient or, after protest, when ordered to so

by the judge. See W. C. Tait, "The Physician's Obligation to Secrecy," *American Medicine* 4 (1902): 267.

96. Dr. Krahmer, "Der Zeugnisszwang der Aerzte," *Berliner Klinische Wochenschrift* 1875, nos. 26–27: 365–66, 378.

97. Liebmann, *Die Pflicht des Arztes*, 6–7; Siegfried Placzek, *Das Berufsgeheimnis des Arztes* (Leipzig: Verlag von Georg Thieme, 1893), 6.

98. Liebmann, *Die Pflicht des Arztes*, 7; Simonson, "Das Berufsgeheimnis der Aerzte und deren Recht der Zeugnisverweigerung," *Deutsche Juristen-Zeitung* 9 (1904): 1014–17. Likewise, see Josef Kohler, "Stellung der Rechtsordnung zur Gefahr der Geschlechtskrankheiten," *Zeitschrift für Bekämpfung der Geschlechtskrankheiten* 2 (1903/4): 30; Hugo Sandheim, "Die unbefugte Offenbarung von Privatgeheimnissen nach § 300 St. G. B." (Jur. Diss., Friedrichs-Universität Halle-Wittenberg, 1904), 29; Schmidt, *Berufsgeheimnis*, 21–23; Exner, "Das Berufsgeheimnis," 42–45.

99. Justus Olshausen, *Kommentar zum Strafgesetzbuch für das Deutsche Reich*, 4th revised edition, vol. 2 (Berlin: Verlag von Franz Vahlen, 1892), 1142; Fromme, *Die rechtliche Stellung des Arztes und seine Pflicht zur Verschwiegenheit im Beruf* (= *Berliner Klinik*, No. 165) (Berlin: Fischer, 1902), 20–22; Jummel, "Der § 300 Str.G.B.," 46; Walther Zschok, "§ 300 StrGB" (Jur. Diss., Universität Rostock, 1903), 46; Léon Seréxhe, "Die Verletzung fremder Geheimnisse" (Jur. Diss., Universität Freiburg i. B., 1906), 84–86; Richard Otto Gans, "Das ärztliche Berufsgeheimnis des § 300 RStrGB" (Jur. Diss., Ruprecht-Karls-Universität Heidelberg, 1907), 27–29.

100. *Entscheidungen des Reichsgerichts in Strafsachen* 19 (1889): 364–67 (decision of July 8, 1889).

101. Placzek, *Berufsgeheimnis* (1893), 6–8; Dr. Schlegtendal, "Das Berufsgeheimniss der Aerzte," *Deutsche Medicinische Wochenschrift* 1895, No. 31: 503–4.

102. Dr. Marcus, "Wegen Verweigerung des Zeugnisses," *Aerztliches Vereinsblatt* 11 (1884): 94–96.

103. Dr. jur. Biberfeld, "Die Schweigepflicht des Arztes," *Zeitschrift für Medizinal-Beamte* 15 (1902): 648–50.

104. Dr. H., "Ueber Verschwiegenheitspflicht und Zeugnisverweigerungsrecht des Arztes vor Gericht," *Aerztliches Vereinsblatt* 30 (1903): 248–51.

105. *Entscheidungen des Reichsgerichts in Civilsachen* 53 (1903): 315–19 (decision of January 19, 1903).

106. Ibid., 317–18. This decision had consequences for later cases dealing with breach of medical secrecy. See chapter 2, under "A Landmark Decision in Germany, Its Context, and Its Consequences."

107. Siegfried Placzek, "Aerztliches Berufsgeheimnis und Ehe," in *Krankheiten und Ehe. Darstellung der Beziehungen zwischen Gesundheits-Störungen und Ehegemeinschaft*, ed. H. Senator and S. Kaminer (Munich: J. F. Lehmann, 1904), 800–803.

108. John Braxton Hicks, "Notes of Cases in Obstetric Jurisprudence," *The Lancet* 126 (1885): 285.

109. John Glaister, *A Text-Book of Medical Jurisprudence and Toxicology* [1st ed. 1902]. 4th ed. (Edinburgh: E. & S. Livingstone, 1921), 58; Robert Saundby, *Medical Ethics: A Guide to Professional Conduct*, 2nd ed. (London: Charles Griffin & Company, 1907), 42. See also "Foreign Letters: London: Medical Confidences," *JAMA* 89 (1927): 1346.

110. Percy Clarke and Charles Meymott Tidy, *Medical Law for Medical Men: Their Legal Relations Shortly and Popularly Explained* (London: Baillière, Tindall, and Cox, 1890), 39.

111. "Kitson v. Playfair and Wife," *The Times*, March 26 and 28, 1896. For details of this case, see Angus McLaren, "Privileged Communications: Medical Confidentiality in Late Victorian Britain," *Medical History* 37 (1993): 129–47, and chapter 3, under "'A Monstrous Cruelty': The *Kitson v. Playfair* Trial in London and Its Impact."

112. "The Seal of Professional Secrecy," *The Lancet* 155 (1900): 1292–93. See also Andreas-Holger Maehle and Sebastian Pranghofer, "Medical Confidentiality in the Late Nineteenth and Early Twentieth Centuries: An Anglo-German Comparison," *Medizinhistorisches Journal* 45 (2010): 197. Breach of confidence was actually a rather rare offense according to the official Minutes of the General Medical Council. See Russell G. Smith, *Medical Discipline. The Professional Conduct Jurisdiction of the General Medical Council, 1858–1990* (Oxford: Clarendon Press, 1994), 103, 106.

113. Placzek, *Berufsgeheimnis* (1909), 81.

114. Muir Mackenzie, "Memorandum on the Law of the Obligation of Medical Practitioners with Regard to Professional Secrecy," *The Lancet* 153 (1899): 787–88.

115. Mackenzie, "Memorandum," 788. On the libel charge against Playfair by his sister-in-law, Linda Kitson, and his alleged breach of confidence, see McLaren, "Privileged Communications" and chapter 3, under "'A Monstrous Cruelty': The *Kitson v. Playfair* Trial in London and Its Impact."

116. Mackenzie, "Memorandum," 787.

117. Cf. McLaren, "Privileged Communications," 137, and chapter 3, under "'A Monstrous Cruelty': The *Kitson v. Playfair* Trial in London and Its Impact."

118. Robert Brudenell Carter, *Doctors and Their Work or Medicine, Quackery, and Disease* (London: Smith, Elder, and Co., 1903), 277.

119. See chapter 2, under "British Discussions on Medical Secrecy and Venereal Diseases."

120. "Law Report, Jan. 13," *The Times*, January 14, 1920, 5; "Medical Evidence in Divorce. Effect on Treatment Schemes. Value of Secrecy," *The Times*, January 15, 1920; Digby Cotes-Preedy, "Judges and Medical Privilege," *The*

Times, January 15, 1920, 5; "The Privacy of Venereal Clinics," *The Lancet* 195 (1920): 163; D. Harcourt Kitchin, *Law for the Medical Practitioner* (London: Eyre & Spottiswoode Ltd., 1941), 160–61; Angus H. Ferguson, "Speaking Out about Staying Silent: An Historical Examination of Medico-Legal Debates over the Boundaries of Medical Confidentiality," in *Lawyers' Medicine: The Legislature, the Courts and Medical Practice, 1760–2000*, ed. Imogen Goold and Catherine Kelly (Oxford: Hart, 2009), 111–12.

121. "A Doctor's Claim to Privilege. Needham v. Needham and Bennett," *The Times*, June 10, 1921, 4; "Dr. John Elliott," *The Times*, December 20, 1921, 12; Ferguson, "Speaking Out," 114–17.

122. For a detailed discussion of the position of the Ministry of Health, see Ferguson, *Should a Doctor Tell?*, 55–77.

123. Ferguson, "Speaking Out," 118–19. For the extensive discussions of the British Medical Association on confidentiality and medical privilege, see Ferguson, *Should a Doctor Tell?*, 79–109; and Andrew A. G. Morrice, "'Should the Doctor Tell?' Medical Secrecy in Early Twentieth-Century Britain," in *Medicine, Health and the Public Sphere in Britain, 1600–2000*, ed. Steve Sturdy (London: Routledge, 2002), 60–82.

124. Viscount Birkenhead, *Points of View* (London: Hodder and Stoughton Limited, 1922), vol. 1, 75. For further details and background to the Lord Chancellor's position in the issue of a medical privilege, see Ferguson, *Should a Doctor Tell?*, 111–23.

125. William Sanderson and E. B. A. Rayner, *An Introduction to the Law and Tradition of Medical Practice* (London: H. K. Lewis & Co., 1926), 51.

126. "Professional Secrecy of Doctors. Mr. Justice McCardie on Duty to Court," *The Times*, July 19, 1927, 11; E. Graham Little, "Medical Privilege. Doctors and the Courts," *The Times*, November 14, 1927, 10; Ferguson, *Should a Doctor Tell?*, 130–31.

127. "Foreign Letters: London: Law and Medicine," *JAMA* 90 (1928): 629.

128. The widening of the bill's scope followed the argument that a restriction to VD cases alone would indirectly confirm the disease whenever medical privilege was claimed. For details on Graham-Little's bills and their debate, see Ferguson, *Should a Doctor Tell?*, 138–53. Graham-Little's initiatives were also reported in the United States: "Foreign Letters: London: Medical Secrecy," *JAMA* 90 (1928): 43; "Foreign Letters: London: Rejection of Bill Concerning Professional Secrets," *JAMA* 108 (1937): 983.

129. Sabine Michalowski, *Medical Confidentiality and Crime* (Aldershot: Ashgate, 2003), 168.

130. This applied also to Scotland, where the precedent of the Duchess of Kingston case was likewise influential. See Ferguson, *Should a Doctor Tell?*, 25–26.

131. The Medicolegal Bureau of the American Medical Association advised

in 1914, "Because of the varying provisions of the law in different states, it is important that physicians do not take too much for granted, but that they secure legal advice, especially as to the law in their respective states." "Privileged Communications," *JAMA* 62 (1914): 1351. See also, from the legal point of view, E. B. P., "Evidence: Privileged Communications to Physicians: Waiver," *California Law Review* 6 (1918): 300–302. For an overview of the numerous different rules and decisions regarding medical privilege in the various American states up to 1930, see Alfred W. Herzog, *Medical Jurisprudence* (Indianapolis: Bobbs-Merrill, 1931), 117–38.

Chapter Two

1. The Wassermann test, a blood serum test for syphilis, had been developed in 1906 by the Berlin bacteriologist Paul August von Wassermann in collaboration with Albert Neisser and Carl Bruck.

2. "Disclosure of Confidential Information as to Contagious Disease (Simonsen v. Swenson (Neb.), 177 N. W. R. 831)," *Journal of the American Medical Association* 75 (1920): 1153.

3. Ibid., 1154.

4. Ibid.

5. "The Physician's Responsibility: An Important Decision on Professional Secrecy" (Editorial), *JAMA* 75 (1920): 1207–8; "Liability of a Physician for Revealing out of Court his Patient's Confidences," *Harvard Law Review* 34 (1920–21): 312–14; "Privileged Communications to Physicians," *Yale Law Journal* 30 (1921): 289–91. See also Samuel I. Jacobs, "Evidence: Privileged Communications between Physician and Patient in California: Cal. Code Civ. Proc. § 1881 (4)," *California Law Review* 20 (1932): 312.

6. "Liability of a Physician," 314.

7. "Privileged Communications to Physicians," 291.

8. "The Physician's Responsibility," 1207.

9. Alan Brandt, *No Magic Bullet: A Social History of Venereal Disease in the United States since 1880* (New York: Oxford University Press, 1985), 11–12, 40–41.

10. Amy L. Fairchild, Ronald Bayer, and James Colgrove, with Daniel Wolfe, *Searching Eyes: Privacy, the State, and Disease Surveillance in America* (Berkeley: University of California Press, 2007), 64–66.

11. See also Brandt, *No Magic Bullet*, 42–43.

12. On the history of tuberculosis reporting and surveillance in the United States, see Fairchild et al., *Searching Eyes*, 33–57.

13. William Edler, "The Reporting of Venereal Diseases by Physicians," *JAMA* 74 (1920): 1764.

14. Ibid., 1765.

15. Brandt, *No Magic Bullet*, 7–11, 14–17, 19–20.

16. Prince A. Morrow, *Social Diseases and Marriage: Social Prophylaxis* (New York: Lea Brothers & Co., 1904), 46.

17. Ibid., 50–51.

18. Brandt, *No Magic Bullet*, 11. To protect the future wife and children, Fournier demanded that a syphilis-infected man should delay marriage for three to four years and complete a course of treatment. Annet Mooij, *Out of Otherness: Characters and Narrators in the Dutch Venereal Disease Debates 1850–1990* (Amsterdam: Rodopi, 1998), 49.

19. Morrow, *Social Diseases*, 53–59. See Georges Thibierge, *Syphilis et Déontologie* (Paris: Masson et Cie, 1903).

20. Morrow, *Social Diseases*, 61.

21. Ibid., 62–63. See Paul Brouardel, *Le Secret Médical* (Paris: J.-B. Baillière et Fils, 1887), 49–50; Paul Brouardel, *La Responsabilité Médicale* (Paris: J.-B. Baillière et Fils, 1898), 146.

22. Morrow, *Social Diseases*, 63–65.

23. Charlotte Perkins Gilman, *The Crux*, edited and with an introduction by Jennifer S. Tuttle (Newark: University of Delaware Press, 2002), 185–86.

24. Ibid., 55, 191, 232.

25. Cf. ibid. (introduction by Tuttle), 26, 30–33.

26. Morrow, *Social Diseases*, 66–69.

27. William Archer Purrington, "Professional Secrecy and the Obligatory Notification of Venereal Diseases," *New York Medical Journal* 85 (1907): 1206–10.

28. See chapters 1 and 3.

29. "Medical Confidences and Professional Honor," *JAMA* 26 (1896): 784.

30. "Medical Confidences and Medical Testimony," *JAMA* 31 (1898): 309.

31. Ibid.

32. Brandt, *No Magic Bullet*, 19; "Michigan Medical Laws," *JAMA* 33 (1899): 233.

33. "Michigan Medical Laws," *JAMA* 33 (1899): 233; "Ethics of Professional Secrets," *JAMA* 35 (1900): 363.

34. "Regulation of Venereal Disease," *JAMA* 39 (1902): 776; A. L. Benedict, "The Limitation of the Principle of Privileged Communications," *American Medicine* 10 (1905): 705; Frances M. Greene, "Reportability of Syphilis and Gonorrhea," *JAMA* 57 (1911): 1051–52; John N. Hurty, "The President's Address," *American Journal of Public Health* 2 (1912): 760.

35. Ibid., 760–61.

36. Norman Barnesby, *Medical Chaos and Crime* (London: Mitchell Kennerley, 1910), 210–18. On Barnesby and his criticisms of contemporary medical and surgical practice, see Peter J. Kernahan, "'A Condition of Development': Muckrakers, Surgeons, and Hospitals, 1890–1920," *Journal of the American College of Surgeons* 206 (2008): 376–84.

37. Judith Walzer Leavitt, *Typhoid Mary: Captive to the Public's Health* (Boston: Beacon Press, 1996).

38. Robert Baker, *Before Bioethics: A History of American Medical Ethics from the Colonial Period to the Bioethics Revolution* (New York: Oxford University Press, 2013), 219–20.

39. American Medical Association, "Principles of Medical Ethics (1912)," in *The American Medical Ethics Revolution. How the AMA's Code of Ethics Has Transformed Physicians' Relationships to Patients, Professionals, and Society*, ed. Robert B. Baker, Arthur L. Caplan, Linda L. Emanuel, and Stephen R. Latham (Baltimore: Johns Hopkins University Press, 1999), 354.

40. Ibid., 346.

41. Baker, *Before Bioethics*, 222.

42. AMA, "Principles of Medical Ethics (1912)," 346.

43. Mark Thomas Connelly, "Prostitution, Venereal Disease, and American Medicine," in *Women and Health in America: Historical Readings*, ed. Judith Walzer Leavitt (Madison: University of Wisconsin Press, 1984), 203–4, 217–18; Brandt, *No Magic Bullet*, 19–20, 147.

44. Ibid., 147–49.

45. "Infringement of Professional Secrecy," *JAMA* 44 (1905): 803; "Professional Secrecy," *JAMA* 44 (1905): 1865.

46. *Entscheidungen des Reichsgerichts in Strafsachen* 38 (1905): 62–63, 65 (decision of May 16, 1905).

47. Albert Moll, *Ärztliche Ethik: Die Pflichten des Arztes in allen Beziehungen seiner Thätigkeit* (Stuttgart: Ferdinand Enke, 1902). For a discussion of Moll's medical ethics, see Andreas-Holger Maehle, "'God's Ethicist': Albert Moll and His Medical Ethics in Theory and Practice," *Medical History* 56 (2012): 217–36.

48. Albert Moll, *Ein Leben als Arzt der Seele: Erinnerungen* (Dresden: Carl Reissner Verlag, 1936), 270; *Entscheidungen des Reichsgerichts in Civilsachen* 53 (1903): 317 (decision of January 19, 1903).

49. Moll, *Ein Leben*, 271; Fromme, *Die rechtliche Stellung des Arztes und seine Pflicht zur Verschwiegenheit im Beruf* (= *Berliner Klinik*, No. 165) (Berlin: Fischer, 1902), 27–28. Friedrich Ottomar Jummel ("Der § 300 Str.G.B., ein Versuch seiner Auslegung" [Jur. Diss., Universität Leipzig, 1903], 51–53) shared Fromme's view and further supported it with the Supreme Court decision of January 19, 1903. Jummel saw an entitlement especially of family doctors to warn other members of the family or household in case of contagious diseases. The same position was taken by Léon Seréxhe, "Die Verletzung fremder Geheimnisse" (Jur. Diss., Universität Freiburg i. B., 1906), 93–95. See also Andreas-Holger Maehle, "Protecting Patient Privacy or Serving Public Interests? Challenges to Medical Confidentiality in Imperial Germany," *Social History of Medicine* 16 (2003): 392.

50. *Entscheidungen des Reichsgerichts in Strafsachen* 38 (1905): 64–66.

51. Hugo Heinemann, "Das ärztliche Berufsgeheimnis," *Deutsche Medizinische Wochenschrift* 26 (1905): 1037–38; "'Befugte' Offenbarung eines Privatgeheimnisses seitens eines Arztes (§ 300 Str.-G.-B.)," *Aerztliches Vereinsblatt für Deutschland* 32 (1905): 556–58; "Berlin. The Law as to Professional Secrecy," *British Medical Journal* 1905 (June 2): 1316.

52. Moll, *Ein Leben*, 271.

53. F. Ottmer, *Schweigen. Erzählung* (Berlin: Concordia Deutsche Verlags-Anstalt, 1902).

54. Hoerning, "Berufsgeheimniss," *Deutsche Medicinische Wochenschrift* 23 (1903): 415–16.

55. Albert Hellwig, "Die civilrechtliche Bedeutung der Geschlechtskrankheiten," *Zeitschrift für Bekämpfung der Geschlechtskrankheiten* 1 (1903): 33–34.

56. Siegfried Placzek, "Aerztliches Berufsgeheimnis und Ehe," in *Krankheiten und Ehe. Darstellung der Beziehungen zwischen Gesundheits-Störungen und Ehegemeinschaft*, ed. H. Senator and S. Kaminer (Munich: J. F. Lehmann, 1904), 798. An abridged American edition of this volume (New York: Paul B. Hoeber, 1909) was briefly reviewed in *JAMA* 52 (1909): 2129. There was also an abridged English edition (London: Rebman Limited, 1907). Placzek's book *Das Berufsgeheimnis des Arztes* (Leipzig: Georg Thieme) went through three editions (1893, 1898, and 1909).

57. Wilhelm Rudeck, *Medizin und Recht: Geschlechtsleben und -Krankheiten in medizinisch-juristisch-kulturgeschichtlicher Bedeutung*, 2nd ed. (Berlin: H. Barsdorf, 1902), 10. See also S. Placzek, "Medico-Professional Secrecy in Relation to Marriage," in *Marriage and Disease. Being an Abridged Edition of "Health and Disease in Relation to Marriage and the Married State,"* ed. H. Senator and S. Kaminer (London: Rebman Limited, 1907), 438–39.

58. See, for example, Walther Zschok, "§ 300 StrGB" (Jur. Diss., Universität Rostock, 1903), 55; Martin Chotzen, "Meldepflicht und Verschwiegenheitspflicht des Arztes bei Geschlechtskrankheiten," *Zeitschrift für Bekämpfung der Geschlechtskrankheiten* 2 (1903/4): 451. The legal argument that a syphilitic groom requiring secrecy from his physician thereby turned the doctor–patient relationship into an immoral, invalid contract, and that the physician could therefore warn the bride, does not seem to have had much impact. Cf. Walter Jellinek, "Der Umfang der Verschwiegenheitspflicht des Arztes und des Anwalts," *Monatsschrift für Kriminalpsychologie und Strafrechtsreform* 3 (1906/7): 685; Bjarne Exner, "Das Berufsgeheimnis des Arztes gemäß § 300 des Str. G. B." (Jur. Diss., Ruprecht-Karls-Universität Heidelberg, 1909), 52–53.

59. See, for example, Ernst Friedersdorff, "Die unbefugte Offenbarung von Privatgeheimnissen durch Rechtsbeistände, Medizinalpersonen und ihre Gehülfen. § 300 St.G.B." (Jur. Diss., Friedrichs-Universität Halle-Wittenberg, 1906), 31–33, 49; Richard Otto Gans, "Das ärztliche Berufsgeheimnis des § 300

RStrGB" (Jur. Diss., Ruprecht-Karls-Universität Heidelberg, 1907), 32–35; Exner, "Das Berufsgeheimnis," 47–49; Hans Weizmann, "Das Berufsgeheimnis. (§ 300 RStGB.)" (Jur. Diss., Universität Breslau, 1909), 29–30.

60. Pallaske, "Die Schweigepflicht des Arztes," *Deutsche Juristen-Zeitung* 11 (1906): 295–96; Heinrich Schmidt, *Das ärztliche Berufsgeheimnis* (Jur. Diss., Universität Leipzig, Jena: Gustav Fischer, 1907), 20–21. For a legal defense of the decision, see, however, Jaeger, "Das Berufsgeheimnis der Aerzte und Anwälte," *Deutsche Juristen-Zeitung* 11 (1906): 800–805.

61. Max Alsberg, "Das ärztliche Berufsgeheimnis," *Deutsche Medizinische Wochenschrift* 31 (1908): 1359.

62. Ludwig Bendix, "Zur Verschwiegenheitspflicht der Ärzte," *Zeitschrift für Bekämpfung der Geschlechtskrankheiten* 5 (1906): 375–76.

63. "Durch die Allerhöchste Cabinets-Ordre vom 8. August 1835 genehmigtes Regulativ über die sanitäts-polizeilichen Vorschriften bei den am häufigsten vorkommenden ansteckenden Krankheiten," in F. L. Augustin, *Die Königlich preußische Medicinalverfassung oder vollständige Darstellung aller, das Medicinalwesen und die medicinische Polizei in den Königlich Preußischen Staaten betreffenden Gesetze, Verordnungen und Einrichtungen*, vol. 6: *Medicinalverordnungen 1833 bis 1837* (Potsdam 1838), 978–79; Siegfried Placzek, *Das Berufsgeheimnis des Arztes*, 3rd enlarged and revised edition (Leipzig: Verlag von Georg Thieme, 1909), 157.

64. Ibid., 157, 159–60; Lutz D. H. Sauerteig, *Krankheit, Sexualität, Gesellschaft: Geschlechtskrankheiten und Gesundheitspolitik in Deutschland im 19. und frühen 20. Jahrhundert* (Stuttgart: Franz Steiner Verlag, 1999), 320–21.

65. "Entwurf eines Gesetzes, betreffend die Bekämpfung gemeingefährlicher Krankheiten," in *Stenographische Berichte des Deutschen Reichstages, 10. Legislaturperiode, 1. Session, 1898–1900*, supplement vol. 6: 4189–93.

66. Exner, "Das Berufsgeheimnis," 49–50.

67. Placzek, *Berufsgeheimnis* (1909), 159; Peter Baldwin, *Contagion and the State in Europe, 1830–1930* (Cambridge: Cambridge University Press, 2005), 453; Sauerteig, *Krankheit*, 327.

68. *Stenographische Berichte über die Verhandlungen des Preußischen Hauses der Abgeordneten, 20. Legislaturperiode, I. Session 1904/05* (Berlin: W. Moeser, 1905), vol. 6, 8556–75 (session of January 18, 1905) and vol. 8, 12688–99 (session of April 7, 1905); Heinrich Dittenberger, "Zum § 300 des Reichsstrafgesetzbuches," *Monatsschrift für Kriminalpsychologie und Strafrechtsreform* 2 (1905/6): 54–58; Pallaske, "Die Schweigepflicht," 296–97.

69. Chotzen, "Meldepflicht," 453; Sauerteig, *Krankheit*, 322. See also Lutz D. H. Sauerteig, "'The Fatherland Is in Danger, Save the Fatherland!' Venereal Disease, Sexuality and Gender in Imperial and Weimar Germany," in *Sex, Sin and Suffering: Venereal Disease in European Society since 1870*, ed. Roger Davidson and Lesley A. Hall (London: Routledge, 2001), 84.

70. Ibid., 77.

71. "Second German Preventive Congress," *JAMA* 44 (1905): 1207.

72. Max Flesch, "Das ärztliche Berufsgeheimnis und die Bekämpfung der Geschlechtskrankheiten," *Zeitschrift für Bekämpfung der Geschlechtskrankheiten* 4 (1905): 32–51.

73. Albert Neisser, "Abänderung des § 300 des Reichs-Strafgesetzbuches und ärztliches Anzeigerecht in ihrer Bedeutung für die Bekämpfung der Geschlechtskrankheiten," *Zeitschrift für Bekämpfung der Geschlechtskrankheiten* 4 (1905): 1–28.

74. Max Bernstein, "Ärztliches Berufsgeheimnis und Geschlechtskrankheiten," *Zeitschrift für Bekämpfung der Geschlechtskrankheiten* 4 (1905): 29–31.

75. *Stenographische Berichte über die Verhandlungen des Preußischen Herrenhauses in der Session 1905/06* (Berlin: Julius Sittenfeld, 1906), 265.

76. Ibid., 265–66.

77. *Entscheidungen des Preußischen Ehrengerichtshofes für Ärzte*, vol. 1 (Berlin: Richard Schoetz, 1908), 93–98 (decision of September 27, 1907). See also "Entscheidungen des ärztlichen Ehrengerichtshofs. Die ärztliche Schweigepflicht, Geschlechtskrankheiten," *Berliner Aerzte-Correspondenz* 13 (1908): 55–56. On the development and operation of the Prussian system of medical disciplinary tribunals, the so-called medical courts of honor (*ärztliche Ehrengerichte*), see Andreas-Holger Maehle, *Doctors, Honour and the Law: Medical Ethics in Imperial Germany* (Basingstoke: Palgrave Macmillan, 2009), 6–46.

78. Vollmann, "Umschau. Beratungstellen—Berufsgeheimnis—Kurpfuscherei—ärztliche Ausbildung," *Ärztliches Vereinsblatt für Deutschland* 43 (1916): 458.

79. Sauerteig, *Krankheit*, 337.

80. Sauerteig, "Venereal Disease," 85–86.

81. *Entscheidungen des Preußischen Ehrengerichtshofes für Ärzte*, vol. 4 (Berlin: Richard Schoetz, 1927), 55–56 (decision of February 14, 1925).

82. Rudolf Lehmann, "Die Schweigepflicht des Arztes und das Gesetz zur Bekämpfung der Geschlechtskrankheiten," *Die Medizinische Welt* 20 (1928): 776–78; Walter Spengemann, "Das Berufsgeheimnis des Arztes" (Med. Diss., Westfälische Wilhelms-Universität zu Münster, 1928), 22–23; Sauerteig, "Venereal Disease," 86. Soviet Russia adopted a similar policy with a government decree in 1927 that authorized the health organs to forcibly examine and treat suspected VD patients if they resisted undergoing these procedures voluntarily. The principle of medical secrecy was preserved, however, despite calls to abolish it as superfluous in Soviet medicine where illness should be seen as a misfortune, not a disgrace. Cf. Frances L. Bernstein, "Behind the Closed Door: VD and Medical Secrecy in Early Soviet Medicine," in *Soviet Medicine: Culture, Practice, and Science*, ed. F. L. Bernstein, Christopher Burton, and Dan Healy (DeKalb: Northern Illinois University Press, 2010), 92–110.

83. Lehmann, "Schweigepflicht," 776. However, criticism of the Supreme Court decisions of 1903 and 1905, and their recognition of a doctor's collision of duties, continued as well; see Wilhelm Hilgers, "Schweigepflicht und Zeugnisverweigerungsrecht des Arztes" (Jur. Diss., Rheinische Friedrich-Wilhelms-Universität Bonn, 1930), 29–31.

84. "Berlin. (From Our Own Correspondent.) Control of Venereal Disease," *The Lancet* 1928 (October 27): 889.

85. Sauerteig, "Venereal Disease," 87.

86. Lesley A. Hall, "Venereal Diseases and Society in Britain, from the Contagious Diseases Acts to the National Health Service," in *Sex, Sin and Suffering: Venereal Disease in European Society since 1870*, ed. Roger Davidson and Lesley A. Hall (London: Routledge, 2001), 120–36; Roger Davidson and Lutz D. H. Sauerteig, "Law, Medicine and Morality: A Comparative View of Twentieth-Century Sexually Transmitted Disease Controls," in *Coping with Sickness: Medicine, Law and Human Rights—Historical Perspectives*, ed. John Woodward and Robert Jütte (Sheffield: European Association for the History of Medicine and Health Publications, 2000), 127–47.

87. Graham Mooney, "Public Health versus Private Practice: The Contested Development of Compulsory Infectious Disease Notification in Late-Nineteenth-Century Britain," *Bulletin of the History of Medicine* 73 (1999): 238–67.

88. Robert Saundby, *Medical Ethics: A Guide to Professional Conduct*, 2nd ed. (London: Charles Griffin & Company, 1907), 68, 116.

89. Campbell Williams, "A Lecture on the Ethics of the Medical Profession in Relation to Syphilis and Gonorrhoea. Delivered before the Harveian Society on Feb. 8th, 1906," *The Lancet* 167 (1906): 361–63.

90. See ibid., 546–47, 626–27,789–90, 931–32, 991–92.

91. Williams, "Lecture," 361–62.

92. J. Howell Evans, "The Medico-Legal Significance of Gonorrhoea," *Transactions of the Medico-Legal Society* 5 (1907/8): 157. See, however, the opposite view expressed by F. G. Crookshank, "The Medico-Legal Relations of Venereal Disease," in *The Diagnosis and Treatment of Venereal Diseases in General Practice*, 4th ed., ed. L. W. Harrison (London: Humphrey Milford, Oxford University Press, 1931), 478: "An employer has no sort of claim to obtain from a doctor attending an employee information as to the nature of a malady from which the latter may be suffering, without his express consent: in no case does payment of the doctor's fee, or a promise, express or implied, to pay the fee, give any shadow of right to such information."

93. A. G. Bateman, "Professional Secrecy and Privileged Communications," *Transactions of the Medico-Legal Society* 2 (1904/5): 57–58.

94. Ibid., 58–59.

95. Davidson and Sauerteig, "Law, Medicine and Morality," 131.

96. Ibid., 135–36.

97. This condition was made notifiable under the Public Health (Ophthalmia Neonatorum) Regulations of 1914; see Crookshank, "Medico-Legal Relations of Venereal Disease," 505.

98. Davidson and Sauerteig, "Law, Medicine and Morality," 131–32.

99. A. B. Dunne, "Medical Certificates Prior to Marriage and Venereal Diseases," *The Lancet* 1918 (October 5): 472.

100. "All information obtained in regard to any person treated under a scheme approved in pursuance of this Article shall be regarded as confidential." Cf. Crookshank, "Medico-Legal Relations of Venereal Disease," 504.

101. "Medical Evidence in Divorce. Effect on Treatment Schemes. Value of Secrecy," *The Times*, January 15, 1920, 5.

102. "The Privacy of Venereal Clinics," *The Lancet* 195 (1920): 163. See also C. F. Marshall, "Venereal Disease and Professional Secrecy," *The Lancet* 195 (1920): 171. The *BMJ* took a more positive view of compelling medical witnesses to give evidence in cases involving venereal disease but also thought that legislation on this matter was necessary to avoid a conflict between a doctor's legal and professional duties. See "Medical Professional Privilege," *British Medical Journal* 1920 (January 24): 135–36.

103. "Doctors and Patients," *Daily Chronicle*, June 10, 1921. See also "Doctors Must Tell. Secrecy Pledge Overruled by Judge. No Privilege in Law," *Daily Chronicle*, June 10, 1921.

104. Andreas-Holger Maehle and Sebastian Pranghofer, "Medical Confidentiality in the Late Nineteenth and Early Twentieth Centuries: An Anglo-German Comparison," *Medizinhistorisches Journal* 45 (2010): 198–99; Angus H. Ferguson, *Should a Doctor Tell? The Evolution of Medical Confidentiality in Britain* (Farnham: Ashgate, 2013), 64–68, 111–23. See also chapter 1, above.

105. W. G. Aitchison Robertson, *Medical Conduct and Practice: A Guide to the Ethics of Medicine* (London: A. & C. Black, 1921), 134.

106. Hugh Woods, "Medical Secrecy," *The Lancet* 203 (1924): 853. Wood's article was republished in amended form in *The Conduct of Medical Practice*, ed. by the Editor of "The Lancet" [Samuel Squire Sprigge] and Expert Collaborators (London: The Lancet, 1927), 78–83.

107. Lord Riddell, *Medico-Legal Problems* (London: H. K. Lewis & Co., 1929), 58.

108. Ibid., 54; "An Interesting Case of Medical Secrecy," *The Lancet* 1921 (April 16): 822.

109. Crookshank, "Medico-Legal Relations of Venereal Disease," 460, 464.

110. D. Harcourt Kitchin, *Law for the Medical Practitioner* (London: Eyre & Spottiswoode Ltd., 1941), 282.

111. Cf. Roger Davidson, *Dangerous Liaisons: A Social History of Venereal Disease in Twentieth-Century Scotland* (Amsterdam: Rodopi, 2000), 119–20.

112. Ibid., 297–303.

Chapter Three

1. "Criminal Abortions," *JAMA* 8 (1887): 298.

2. Ibid.

3. James C. Mohr, *Abortion in America: The Origins and Evolution of National Policy, 1800–1900* (New York: Oxford University Press, 1978), 147–70; Leslie J. Reagan, *When Abortion Was a Crime: Women, Medicine, and Law in the United States, 1867–1973* (Berkeley: University of California Press, 1997), 10–14; Robert Baker, *Before Bioethics: A History of American Medical Ethics from the Colonial Period to the Bioethics Revolution* (New York: Oxford University Press, 2013), 174–78.

4. James C. Mohr, "Patterns of Abortion and the Response of American Physicians, 1790–1930," in *Women and Health in America: Historical Readings*, ed. Judith Walzer Leavitt (Madison: University of Wisconsin Press, 1984), 118.

5. Baker, *Before Bioethics*, 171–73, 177–79.

6. Mohr, *Abortion in America*, 160–63.

7. Ibid., 163–64. On the repeals of state medical licensing laws in the 1830s and 1840s, which lifted restrictions on alternative practitioners such as homeopaths and Thomsonians, see Owen Whooley, *Knowledge in the Time of Cholera: The Struggle over American Medicine in the Nineteenth Century* (Chicago: University of Chicago Press, 2013), 59–72. Whooley attributes these repeals in part to a failure of allopathic physicians in addressing the challenge of cholera epidemics.

8. Mohr, *Abortion in America*, 166–70; Mohr, "Patterns of Abortion," 117–19.

9. Edwin M. Hale, *The Great Crime of the Nineteenth Century* (Chicago: C. S. Halsey, 1867); Mohr, *Abortion in America*, 173–76.

10. Hale, *The Great Crime*, 17.

11. Ibid., 20–23.

12. Ibid., 26–27, 31–32, 36.

13. Ibid., 20.

14. Mohr, *Abortion in America*, 200–202.

15. Ibid., vii.

16. Reagan, *When Abortion Was a Crime*, 80–81.

17. Mohr, "Patterns of Abortion," 120–21.

18. Ibid., 121–22.

19. Reagan, *When Abortion Was a Crime*, 80–112.

20. H. C. Markham, "Foeticide and Its Prevention," *JAMA* 9 (1888): 806.

21. Ibid.

22. Charles A. Boston, "The Law of Evidence concerning Confidential Communications between Physician and Patient," in *Medical Jurisprudence, Forensic*

Medicine and Toxicology, ed. R. A. Witthaus and Tracy C. Becker, vol. 1 (New York: William Wood & Company, 1894), 96.

23. Ibid., 102; *People v. Murphy*, 101 N.Y. 126 (1886).

24. Boston, "Law of Evidence," 101.

25. William A. Purrington, "Of Certain Legal Relations of Physicians and Surgeons to Their Patients and to One Another," in *A System of Legal Medicine* (1894), ed. Allan McLane Hamilton and Lawrence Godkin, 2nd ed., vol. 1 (New York: E. B. Treat & Company, 1900), 621–22.

26. *Pierson v. People*, 79 N.Y. 424; Boston, "Law of Evidence," 101–2; Purrington, "Of Certain Legal Relations," 624.

27. *Hewett v. Prime*, 21 Wend. 79 (N.Y. S. Ct. 1839); G. W. Field and John B. Uhle, "Privileged Communications," *The American Law Register* 37 (1889): 11–12, 14; Boston, "Law of Evidence," 125–26.

28. *People v. Brower*, 53 Hun. 217 (1889); 6 N.Y. Supl. 730; William C. Tait, "Professional Secrecy and Its Legal Aspects," *JAMA* 33 (1899): 461–62; Tait, "The Physician's Obligation to Secrecy," *American Medicine* 4 (1902): 267; Boston, "Law of Evidence," 102; Purrington, "Of Certain Legal Relations," 624–25.

29. "Medical Confidences," *JAMA* 33 (1899): 1431.

30. William C. Woodward, "A Brief Statement of the Principles Underlying the Physician's Obligation to Secrecy," *American Medicine* 4 (1902): 739–40.

31. Tait, "The Physician's Obligation," 267.

32. Allen H. Seaman, in "Symposium. Criminal Abortion. The Colorado Law on Abortion," *JAMA* 40 (1903): 1096–98.

33. Charles R. Brock, in "Symposium," 1098.

34. Edmund J. A. Rogers, in "Symposium," 1098; H. G. Wetherill, in "Symposium," 1098–99.

35. H. H. Hawkins, in "Symposium," 1099.

36. Myer Solis Cohen, "The Notification to the Health Authorities of Cases of Abortion and Miscarriage," *JAMA* 53 (1909): 2153–55.

37. Norman Barnesby, *Medical Chaos and Crime* (London: Mitchell Kennerley, 1910), 212–13. Barnesby followed an antiabortion line in his book, complaining about the "nefarious trade" of the abortionists, who were "corrupting what was once an honorable profession." See ibid., 219.

38. Ibid., 230.

39. "Dying Declarations Made after Refusal of Physician to Treat Abortion Case without History—Construction of Amended Statutes with Reference to Privilege and Competency of Witnesses (State vs. Law (Wis.), 136 N. W. R. 803)," *JAMA* 60 (1913): 1829–30.

40. Reagan, *When Abortion Was a Crime*, 113–26.

41. "Kitson v. Playfair and Wife," *The Times*, March 23, 25, 26, and 28, 1896; "Medical Trial. Queen's Bench Division, High Court of Justice. (Before

Mr. Justice Hawkins.) Kitson v. Playfair and Wife," *The Lancet* 147 (1896): 896–97, 961–62; "Kitson v. Playfair and Wife," *British Medical Journal* 1896: 815–19 (March 28), 882–84 (April 4).

42. Angus McLaren, "Privileged Communications: Medical Confidentiality in Late Victorian Britain," *Medical History* 37 (1993): 129–47.

43. "Kitson v. Playfair and Wife," *The Times*, March 26, 1896.

44. Ibid.

45. Ibid.

46. Ibid.

47. "Kitson v. Playfair and Wife," *The Times*, March 28, 1896; "Kitson v. Playfair and Wife," *British Medical Journal* 1896: 883.

48. See later in this same section.

49. "The Action of Kitson v. Playfair and Wife," *The Times*, March 28, 1896.

50. "Privilege—a Medical Difficulty," *The Times*, April 4, 1896; "Medical Privilege," *The Times*, April 7, 1896.

51. "Kitson v. Playfair," *The Lancet* 147 (1896): 1292–93.

52. "Professional Confidences," *The Lancet* 147 (1896): 1524.

53. "Professional Secrecy," *The Lancet* 148 (1896): 254.

54. "The Obligation of Professional Secrecy," *British Medical Journal* 1896: 861 (April 4).

55. Ibid., 862.

56. Ibid.

57. "Some Pathological Features," *British Medical Journal* 1896: 870 (April 4); Clement Godson, "Retained Placenta," *British Medical Journal* 1896: 873 (April 4).

58. "The Obligation of Professional Secrecy. Medical and Legal Opinions. I.—The Legal Issues. [By a Legal Correspondent.]," *British Medical Journal* 1896: 869–70 (April 4).

59. M.D., "Professional Secrecy and the Law," *British Medical Journal* 1896: 871–72 (April 4).

60. S.C.S., "Family Duties and Professional Secrecy," *British Medical Journal* 1896: 872–73 (April 4).

61. "The Case of Kitson v. Playfair," *British Medical Journal* 1896: 989–90 (April 18); "Kitson v. Playfair," *British Medical Journal* 1896: 1161–62 (May 9).

62. Ibid., 1161.

63. Ibid., 1162. Similarly, "Secrecy and Privilege," *Medical Press and Circular* 1896: 403 ("If his [i.e., the medical man's] liabilities are to differ from those of the ordinary layman he is entitled to have them clearly defined and to know exactly what they are.")

64. M.D., "Professional Secrecy," 871; A. M. Cooke, *A History of the Royal College of Physicians of London*, vol. 3 (Oxford: Clarendon Press, 1972), 980.

65. Barbara Brookes, *Abortion in England 1900–1967* (London: Croom Helm,

1988), 22–26; John Keown, *Abortion, Doctors and the Law: Some Aspects of the Legal Regulation of Abortion in England from 1803 to 1982* (Cambridge: Cambridge University Press, 1988), 33–35.

66. Cooke, *History of the Royal College*, 980; Angus H. Ferguson, *Should a Doctor Tell? The Evolution of Medical Confidentiality in Britain* (Farnham: Ashgate, 2013), 40–41.

67. Cooke, *History of the Royal College*, 980–81.

68. Cited in ibid., 981.

69. Robert Saundby, *Medical Ethics: A Guide to Professional Conduct*, 2nd ed. (London: Charles Griffin & Company, 1907), 113–14. See also Ferguson, *Should a Doctor Tell?*, 42.

70. "Professional Secrecy," *The Lancet* 181 (1913): 52.

71. Cooke, *History of the Royal College*, 981–82.

72. R v Annie Hodgkiss: Coroner's Depositions, National Archives, Kew, ASSI 13/44, Box 2, Birmingham Autumn Assizes 1914.

73. Ibid., fol. 15; Cooke, *History of the Royal College*, 982–83.

74. Avory, cited in Cooke, *History of the Royal College*, 983. See also "A Judge on Professional Secrecy," *The Lancet* 184 (1914): 1430–31.

75. Ibid.

76. For details of the subsequent discussions with and in the BMA, see Ferguson, *Should a Doctor Tell?*, 43–49.

77. The British Medical Association, Professional Secrecy, Notes and Memoranda by Mr. W. E. Hempson, January 1915. BMA Archives, Wellcome Library, SA/BMA/D170.

78. Extract from Supplementary Report of Council, Supplement, July 3, 1915, Minutes 542 and 550, British Medical Association, Central Ethical Committee, Meeting of November 9, 1920, Professional Secrecy, Memorandum by Deputy Medical Secretary, Appendix I. BMA Archives, Wellcome Library, SA/BMA/D167. See also "London Letter: Professional Secrecy," *JAMA* 65 (1915): 634–35.

79. "London Letter," 635; Cooke, *History of the Royal College*, 984–85.

80. Ferguson, *Should a Doctor Tell?*, 50.

81. "A Judge on Professional Secrecy" (Editorial), *The Lancet* 185 (1915): 28.

82. Ibid.

83. Ibid., 29.

84. Ibid.

85. *Should a Doctor Tell?*, a British Lion Production, dialogue by Edgar Wallace, directed by Manning Haynes [Press kit, London, 1930], British Film Institute.

86. Only exceptionally reporting of an abortion case to the legal authorities without the patient's consent was advocated by a medical commentator, provided there was "conclusive objective evidence of criminal interference." In such a

case, the medical practitioner's duty to the public and himself would override his obligation of secrecy to the patient. Cf. W. H. Willcox, "Criminal Abortion and Professional Secrecy," *The Lancet* 185 (1915): 97–98.

87. Eduard Seidler, "Das 19. Jahrhundert. Zur Vorgeschichte des Paragraphen 218," in *Geschichte der Abtreibung. Von der Antike bis zur Gegenwart*, ed. Robert Jütte (Munich: Verlag C. H. Beck, 1993), 120–39.

88. Dr. Schlegtendal, "Das Berufsgeheiminis der Aerzte," *Deutsche Medicinische Wochenschrift* 21 (1895): 503.

89. Ibid., 504–5.

90. Siegfried Placzek, *Das Berufsgeheimnis des Arztes* (Leipzig: Verlag von Georg Thieme, 1893), 55.

91. Albert Moll, *Ärztliche Ethik: Die Pflichten des Arztes in allen Beziehungen seiner Thätigkeit* (Stuttgart: Ferdinand Enke, 1902), 105–6.

92. Ibid., 259–60.

93. Siegfried Placzek, *Das Berufsgeheimnis des Arztes*, 3rd enlarged and revised edition (Leipzig: Verlag von Georg Thieme, 1909), 151.

94. Ibid., 154.

95. Friedrich Ottomar Jummel, "Der § 300 Str.G.B., ein Versuch seiner Auslegung" (Jur. Diss., Universität Leipzig, 1903), 42; Richard Otto Gans, "Das ärztliche Berufsgeheimnis des § 300 RStrGB" (Jur. Diss., Ruprecht-Karls-Universität Heidelberg, 1907), 25; Richard Schwerdtfeger, "Die Verletzung des Berufsgeheimnisses nach § 300 RStGB" (Jur. Diss., Friedrich-Alexanders-Universität zu Erlangen, 1926), 27; Wilhelm Rudeck, *Medizin und Recht: Geschlechtsleben und -Krankheiten in medizinisch-juristisch-kulturgeschichtlicher Bedeutung*, 2nd ed. (Berlin: H. Barsdorf, 1902), 27–28. See also Nicolas Eichelbrönner, *Die Grenzen der Schweigepflicht des Arztes und seiner berufsmäßig tätigen Gehilfen nach § 203 StGB im Hinblick auf Verhütung und Aufklärung von Straftaten: Anzeigepflichten, Auskunftspflichten und Offenbarungsbefugnisse gegenüber den Strafverfolgungsbehörden* (Jur. Diss., Universität Würzburg, 2001; Frankfurt/Main: Peter Lang, 2001), 24.

96. Rudeck, *Medizin und Recht*, 27; Walther Zschok, "§ 300 StrGB" (Jur. Diss., Universität Rostock, 1903), 40.

97. Gans, "Das ärztliche Berufsgeheimnis," 25.

98. Albert Moll, "Neuere Fragen zum ärztlichen Berufsgeheimnis," *Berliner Aerzte-Correspondenz* 16 (1911): 1–4.

99. *Entscheidungen des Reichsgerichts in Strafsachen* 48 (1914): 269–74 (decision of May 16, 1914).

100. Cornelie Usborne, *The Politics of the Body in Weimar Germany: Women's Reproductive Rights and Duties* (Ann Arbor: University of Michigan Press, 1992), 156–86, 214. See also Cornelie Usborne, *Cultures of Abortion in Weimar Germany* (New York: Berghahn Books, 2007); Atina Grossmann, *Reforming Sex: The Ger-*

man Movement for Birth Control and Abortion Reform, 1920–1950 (New York: Oxford University Press, 1995).

101. Usborne, *Politics of the Body*, 173–201, 215.

102. Gabriele Czarnowski, "Women's Crimes, State Crimes: Abortion in Nazi Germany," in *Gender and Crime in Modern Europe*, ed. Margaret L. Arnot and Cornelie Usborne (London: UCL Press, 1999), 238–56. On the history of eugenics in Germany, see Paul Weindling, *Health, Race and German Politics between National Unification and Nazism, 1870–1945* (Cambridge: Cambridge University Press, 1989); Hans-Walter Schmuhl, *Rassenhygiene, Nationalsozialismus, Euthanasie: Von der Verhütung zur Vernichtung "lebensunwerten Lebens," 1890–1945*, 2nd ed. (Göttingen: Vandenhoeck & Ruprecht, 1992).

103. Wilhelm Schmitz, "Sterilisierungsgesetz und ärztliches Berufsgeheimnis," *Die Medizinische Welt* 9 (1935): 205–6; Udo Benzenhöfer, *Zur Genese des Gesetzes zur Verhütung erbkranken Nachwuchses* (Münster: Klemm & Oelschläger, 2006), 90–93, 117–29. This duty of secrecy was later extended per ministerial decree to include also the personnel of the welfare organizations who had official knowledge of the sterilizations, for example, through assessing how the costs for the operations would be met. See "Runderlaß des Reichs-und Preußischen Ministers des Innern, betr. Schweigepflicht bei Durchführung des Gesetzes zur Verhütung erbkranken Nachwuchses. Vom 31. Mai 1935," *Ärzteblatt für Berlin* 40 (1935): 257. On the other hand, it was suggested that the doctors involved in the sterilizations could be released from medical confidentiality by the chairs of the Hereditary Health Courts so that spouses, relatives, and (in case of children) teachers could be informed about the intervention. Cf. Franz Neukamp (Landgerichtsrat i. R.), "Einige Vorschläge für Erb- und Ehegesundheitssachen," *Monatsschrift für Kriminalpsychologie und Strafrechtsreform* 27 (1936): 250.

104. Ernst Böttger, "Die Grenzen der Schweigepflicht nach § 7 des Gesetzes zur Verhütung erbkranken Nachwuchses vom 14. 7. 1933 (RGBl. I, 529)," *Ärzteblatt für Berlin* 40 (1935): 76.

105. Coermann (Amtsgerichtsrat i. R.), "Gerichtsentscheidungen," *Deutsches Ärzteblatt* 66 (1936): 956.

106. Daniel J. Kevles, *In the Name of Eugenics: Genetics and the Uses of Human Heredity* (Berkeley: University of California Press, 1985), 100. See also Wendy Kline, "Eugenics in the United States," in *The Oxford Handbook of the History of Eugenics*, ed. Alison Bashford and Philippa Levine (Oxford: Oxford University Press, 2010), 511–22.

107. "Human Sterilization in Germany and the United States" (Editorial), *JAMA* 102 (1934): 1501–2.

108. Lucy Bland and Lesley A. Hall, "Eugenics in Britain: The View from the Metropole," in *The Oxford Handbook of the History of Eugenics*, ed. Alison Bashford

and Philippa Levine (Oxford: Oxford University Press, 2010), 219. See also Richard A. Soloway, *Demography and Degeneration: Eugenics and the Declining Birthrate in Twentieth-Century Britain* (Chapel Hill: University of North Carolina Press, 1995).

109. John Macnicol, "Eugenics and the Campaign for Voluntary Sterilization in Britain between the Wars," *Social History of Medicine* 2 (1989): 157–58.

110. Lord Hodder, "An Address on Eugenics and the Doctor," *British Medical Journal* 3805 (1933): 1057–60.

111. Reichsärzteordung (December 13, 1935), § 13 (3). Cf. Albert Hellwig, "Die Neuregelung des Berufsgeheimnisses des Arztes," *Deutsche Medizinische Wochenschrift* 62 (1936): 153–54; Wilhelm Schmitz, "Reichsärzteordnung und Berufsgeheimnis," *Die Medizinische Welt* 10 (1936): 757–58; Oswald Moser, "Das ärztliche Berufsgeheimnis (RAeO. vom 15. XII. 1935)" (Med. Diss., Ernst-Moritz-Arndt-Universität Greifswald, 1936), 8–9.

112. Carl Lübke, "Die Schweigepflicht der Rechtsanwälte, Aerzte und ihrer Gehilfen" (Jur. Diss., Universität Jena, 1920), 45; Schwerdtfeger, "Verletzung des Berufsgeheimnisses," 36–38; Hans Schmalzl, "Der Schutz des Privatgeheimnisses. Ein Vergleich zwischen § 300 St.G.B. und § 293 Entwurf 1925 mit Anhang: Der Schutz des Privatgeheimnisses nach § 325 Entwurf 1927" (Jur. Diss., Friedrich-Alexanders-Universität zu Erlangen, 1928), 13–18, 78–81.

113. Gerhard Gransee, "Das Berufsgeheimnis und sein strafrechtlicher Schutz im geltenden Recht und in den Entwürfen" (Jur. Diss., Universität Leipzig, 1928), 26–27, 88; Willy Schumacher, *Das ärztliche Berufsgeheimnis nach § 300 RStGB* (Berlin: Verlagsbuchhandlung von Richard Schoetz, 1931), 110; Walter Spengemann, "Das Berufsgeheimnis des Arztes" (Med. Diss., Westfälische Wilhelms-Universität zu Münster, 1928), 26; Schmalzl, "Schutz des Privatgeheimnisses," 85.

114. Gerhard Wagner, "Die Reichsärzteordnung ein Instrument nationalsozialistischer Gesundheitspolitik," *Deutsches Ärzteblatt* 65 (1935): 1233–34.

115. See also Eichelbrönner, *Grenzen der Schweigepflicht*, 27–29.

116. Albert Hellwig, "Die Bedeutung der Reichsärzteordnung für die Rechtspflege," *Deutsche Justiz* 98 (1936): 373.

117. Schmitz, "Reichsärzteordnung und Berufsgeheimnis," 758.

118. Wolfgang Mittermaier, "Das ärztliche Berufsgeheimnis nach der Reichsärzteordnung," *Monatsschrift für Kriminalpsychologie und Strafrechtsreform* 27 (1936): 153–54.

119. Eduard Vogler, "Betrachtungen zur Begrenzung der ärztlichen Schweigepflicht nach § 13 Abs. 3 der Reichsärzteordnung" (Jur. Diss., Philipps-Universität zu Marburg, 1939), 32–34. Likewise, see Franz Neukamp (Landgerichtsrat i. R., Bielefeld), "Das ärztliche Berufsgeheimnis nach der Reichsärzteordnung," *Monatsschrift für Kriminalpsychologie und Strafrechtsreform* 28 (1937): 81.

120. Wilhelm Forster, "Das ärztliche Berufsgeheimnis nach dem Recht der Reichsärzteordnung und seine Geltung im Recht der Reichsversicherungsordnung" (Jur. Diss., Leopold-Franzens-Universität zu Innsbruck, 1939), 11, 51.

121. Ibid., 20; Vogler, "Betrachtungen," 36–37; Hellwig, "Neuregelung," 154; Mittermaier, "Das ärztliche Berufsgeheimnis," 154.

122. Ibid.

123. Hans Baruth, "Der Arzt und seine Schweigepflicht" (Med. Diss., Philipps-Universität zu Marburg, 1939), 21–22.

124. See ibid., 22.

125. Christian Müller-Welt, "Das Berufsgeheimnis des Arztes und Apothekers nach der Reichsärzte -und Apothekerordnung" (Jur. Diss., Friedrich-Alexanders-Universität zu Erlangen, 1938), 40–41.

126. Ibid., 42.

127. Hermann Trost, "Das ärztliche Berufsgeheimnis nach § 13 der Reichsärzteordnung" (Jur. Diss., Universität Rostock, 1940), 56–60.

128. Alfred Heger, "Berufsgeheimnis und Abtreibung" (Med. Diss., Julius-Maximilians-Universität Würzburg, 1940), 26.

129. Gustav Aschaffenburg, "Das ärztliche Berufsgeheimnis bei Schwangerschaften," *Monatsschrift für Kriminalpsychologie und Strafrechtsreform* 27 (1936): 155–56.

130. Such regulations existed for Baden-Wurttemberg (1949), Bremen (1949), Hamburg (1949), Hessen (1950), Lower Saxony (1949), North Rhine-Westphalia (1950), Rhineland-Palatinate (1949), and Schleswig-Holstein (1949). See Eichelbrönner, *Grenzen der Schweigepflicht*, 35–37.

131. Ibid.; Hermann Wichmann, *Das Berufsgeheimnis als Grenze des Zeugenbeweises. Ein Beitrag zur Lehre von den Beweisverboten* (Jur. Diss., Universität Göttingen, 2000; Frankfurt/Main: Peter Lang, 2000), 31–32.

132. Ibid., 32–33.

General Conclusions

1. Claudia Huerkamp, *Der Aufstieg der Ärzte im 19. Jahrhundert. Vom gelehrten Stand zum professionellen Experten: Das Beispiel Preußens* (Göttingen: Vandenhoeck & Ruprecht, 1985), 93–94.

2. Catherine Kelly and Imogen Goold, "Introduction: Lawyers' Medicine: The Interaction of the Medical Profession and the Law, 1760–2000," in *Lawyers' Medicine: The Legislature, the Courts and Medical Practice, 1760–2000*, ed. I. Goold and C. Kelly (Oxford: Hart, 2009), 6.

3. Owen Whooley, *Knowledge in the Time of Cholera: The Struggle over American Medicine in the Nineteenth Century* (Chicago: University of Chicago Press, 2013), 68–72, 108.

4. See James C. Mohr, *Licensed to Practice: The Supreme Court Defines the*

American Medical Profession (Baltimore: Johns Hopkins University Press, 2013). The new licensing law of New York State, in 1880, still recognized graduates from homeopathic colleges. Cf. Robert Baker, *Before Bioethics: A History of American Medical Ethics from the Colonial Period to the Bioethics Revolution* (New York: Oxford University Press, 2013), 364–65.

5. Jean McHale, "From *X v Y* to care.data and Beyond: Health Care Confidentiality and Privacy in the C21st: A Critical Turning Point?," *Journal of Medical Law and Ethics* 3 (2015): 109–33; Michael Soljak, "Big Voice or Big Data? The Difficult Birth of care.data," *Journal of Medical Law and Ethics* 3 (2015): 135–42.

6. Laura Donnelly, "Hospital Records of All NHS Patients Sold to Insurers," *The Telegraph*, February 23, 2014, http://www.telegraph.co.uk/news/health/news/10656893/Hospital-records-of-all-NHS-patients-sold-to-insurers.html.

7. The All Party Parliamentary Group for Patient and Public Involvement in Health and Social Care, *Care.data Inquiry* (November 2014), http://patients-association.org.uk/wp-content/uploads/2014/06/APPG-Report-on-Care-data.pdf, 24.

8. Soljak, "Big Voice or Big Data?," 140.

9. Nuffield Council on Bioethics, *The Collection, Linking and Use of Data in Biomedical Research and Health Care: Ethical Issues* (London 2015), http://nuffieldbioethics.org/wp-content/uploads/Biological_and_health_data_web.pdf, 114.

10. Soljak, "Big Voice or Big Data?," 140.

11. McHale, "From X *v* Y to care.data," 133.

Bibliography

Archival Sources

The British Medical Association, Professional Secrecy, Notes and Memoranda by Mr. W. E. Hempson, January 1915, BMA Archives, Wellcome Library, SA/BMA/D170.

Extract from Supplementary Report of Council, Supplement, July 3rd, 1915, Minutes 542 and 550, British Medical Association, Central Ethical Committee, Meeting of November 9th, 1920, Professional Secrecy, Memorandum by Deputy Medical Secretary, Appendix I. BMA Archives, Wellcome Library, SA/BMA/D167.

R v Annie Hodgkiss: Coroner's Depositions, National Archives, Kew, ASSI 13/44, Box 2, Birmingham Autumn Assizes 1914.

Scrapbook for Committee on Ethics, Archive of the New York Academy of Medicine, New York.

Should a Doctor Tell? A British Lion Production. Dialogue by Edgar Wallace. Directed by Manning Haynes [Press kit, London, 1930]. British Film Institute.

Primary Literature

"The Action of Kitson v. Playfair and Wife." *The Times*, March 28, 1896.

All-Party Parliamentary Group for Patient and Public Involvement in Health and Social Care. *Care.data Inquiry* (November 2014), http://patients-association.org.uk/wp-content/uploads/2014/06/APPG-Report-on-Care-data.pdf.

Alsberg, Max. "Das ärztliche Berufsgeheimnis." *Deutsche Medizinische Wochenschrift* 31 (1908): 1356–59.

American Medical Association. "Principles of Medical Ethics (1912)." In *The American Medical Ethics Revolution: How the AMA's Code of Ethics Has Trans-*

formed Physicians' Relationships to Patients, Professionals, and Society, ed. Robert B. Baker, Arthur L. Caplan, Linda L. Emanuel, and Stephen R. Latham, 346–54. Baltimore: Johns Hopkins University Press, 1999.

———. "Proceedings of the Fifty-Fourth Annual Session, Held at New Orleans, May 5, 6, 7 and 8, 1903." *Journal of the American Medical Association* 40 (1903): 1364–86.

Aschaffenburg, Gustav. "Das ärztliche Berufsgeheimnis bei Schwangerschaften." *Monatsschrift für Kriminalpsychologie und Strafrechtsreform* 27 (1936): 155–56.

Augustin, F. L. *Die Königlich preußische Medicinalverfassung oder vollständige Darstellung aller, das Medicinalwesen und die medicinische Polizei in den Königlich Preußischen Staaten betreffenden Gesetze, Verordnungen und Einrichtungen*, vol. 6: *Medicinalverordnungen 1833 bis 1837*. Potsdam, 1838.

Baird, James B. "The Medical Witness. His Rights and Wrongs in Courts of Justice." *Journal of the American Medical Association* 32 (1899): 1148–51.

Barnesby, Norman. *Medical Chaos and Crime*. London: Mitchell Kennerley, 1910.

Baruth, Hans. "Der Arzt und seine Schweigepflicht" (Med. Diss., Philipps-Universität zu Marburg, 1939).

Bateman, A. G. "Professional Secrecy and Privileged Communications." *Transactions of the Medico-Legal Society* 2 (1904/5): 49–75.

Beck, Theodric Romeyn, and John B. Beck. *Elements of Medical Jurisprudence*, 12th ed., revised by C. R. Gilman, vol. 2. Philadelphia: J. B. Lippincott & Co., 1863.

Becker, Tracy C. "Observations Concerning the Law of Privileged Communications between Physician and Patient, as Applicable to the Duties of Railway Surgeons." *Journal of the American Medical Association* 26 (1896): 1065–67.

"'Befugte' Offenbarung eines Privatgeheimnisses seitens eines Arztes (§ 300 Str.-G.-B.)." *Aerztliches Vereinsblatt für Deutschland* 32 (1905): 556–58.

Bendix, Ludwig. "Zur Verschwiegenheitspflicht der Ärzte." *Zeitschrift für Bekämpfung der Geschlechtskrankheiten* 5 (1906): 372–76.

Benedict, A. L. "The Limitation of the Principle of Privileged Communications." *American Medicine* 10 (1905): 703–5.

"Berlin. (From Our Own Correspondent.) Control of Venereal Disease." *The Lancet* 1928 (October 27): 889–90.

"Berlin. The Law as to Professional Secrecy." *British Medical Journal* 1905: 1316 (June 2).

Bernstein, Max. "Ärztliches Berufsgeheimnis und Geschlechtskrankheiten." *Zeitschrift für Bekämpfung der Geschlechtskrankheiten* 4 (1905): 29–31.

Best, William Mawdesley. *The Principles of the Law of Evidence with Elementary Rules for Conducting the Examination and Cross-Examination of Witnesses*, first American, from the sixth London edition of John A. Russell, by James Appleton Morgan, vol. 2. New York: Cockcroft & Company, 1878.

Biberfeld, Dr. jur. "Die Schweigepflicht des Arztes." *Zeitschrift für Medizinal-Beamte* 15 (1902): 648–50.

Birkenhead, Viscount (F. E. Smith). *Points of View*, vol. 1. London: Hodder and Stoughton Limited, 1922.

Boston, Charles A. "The Law of Evidence Concerning Confidential Communications between Physician and Patient." In *Medical Jurisprudence, Forensic Medicine and Toxicology*, ed. R. A. Witthaus and Tracy C. Becker, vol. 1, 89–134. New York: William Wood & Company, 1894.

Böttger, Ernst. "Die Grenzen der Schweigepflicht nach § 7 des Gesetzes zur Verhütung erbkranken Nachwuchses vom 14. 7. 1933 (RGBl. I, 529)." *Ärzteblatt für Berlin* 40 (1935): 76.

Brouardel, Paul. *La Responsabilité Médicale*. Paris: J.-B. Baillière et Fils, 1898.

———. *Le Secret Médical*. Paris: J.-B. Baillière et Fils, 1887.

Brower, Daniel R. "The Necessity of Granting Privileged Communications to the Medical Profession in the State of Illinois." *Journal of the American Medical Association* 27 (1896): 1271–73.

Bundesverfassungsgericht—Entscheidungen, 2 BvR 291/92 (decision of May 22, 2000), http://www.bundesverfassungsgericht.de/entscheidungen/rk20000522_2bvr029192.html.

Busey, Samuel C. "The Code of Ethics." *Journal of the American Medical Association* 34 (1900): 255–59.

Carter, Robert Brudenell. *Doctors and Their Work or Medicine, Quackery, and Disease*. London: Smith, Elder, & Co., 1903.

"The Case of Kitson v. Playfair." *British Medical Journal* 1896: 989–90 (April 18).

Chotzen, Martin. "Meldepflicht und Verschwiegenheitspflicht des Arztes bei Geschlechtskrankheiten." *Zeitschrift für Bekämpfung der Geschlechtskrankheiten* 2 (1903/4): 433–61.

Clarke, Percy, and Charles Meymott Tidy. *Medical Law for Medical Men: Their Legal Relations Shortly and Popularly Explained*. London: Baillière, Tindall, & Cox, 1890.

Coermann (Amtsgerichtsrat i. R.). "Gerichtsentscheidungen." *Deutsches Ärzteblatt* 66 (1936): 956.

Cohen, Myer Solis. "The Notification to the Health Authorities of Cases of Abortion and Miscarriage." *Journal of the American Medical Association* 53 (1909): 2153–55.

"Communications Made to Dentists Are Not Privileged." *Journal of the American Medical Association* 24 (1895): 949–50.

"Construction of Statute as to Privileged Communications and Waivers of Privilege, Especially as between Several Physicians." *Journal of the American Medical Association* 57 (1911): 1158.

"Construction of Statute Relative to Privileged Communications—What Adjudged Waivers—Application to Several Physicians." *Journal of the American Medical Association* 61 (1913): 1837–38.

Cotes-Preedy, Digby. "Judges and Medical Privilege." *The Times*, January 15, 1920.

"Criminal Abortions." *Journal of the American Medical Association* 8 (1887): 298.

Crookshank, F. G. "The Medico-Legal Relations of Venereal Disease." In *The Diagnosis and Treatment of Venereal Diseases in General Practice*, 4th ed., ed. L. W. Harrison, 452–506. London: Humphrey Milford, Oxford University Press, 1931.

Culbertson, Hugh Emmett. *Medical Men and the Law: A Modern Treatise on the Legal Rights, Duties and Liabilities of Physicians and Surgeons*. Philadelphia: Lea & Febiger, 1913.

"Disclosure of Confidential Information as to Contagious Disease (Simonsen v. Swenson (Neb.), 177 N. W. R. 831)." *Journal of the American Medical Association* 75 (1920): 1153–54.

Dittenberger, Heinrich. "Zum § 300 des Reichsstrafgesetzbuches." *Monatsschrift für Kriminalpsychologie und Strafrechtsreform* 2 (1905/6): 54–58.

"A Doctor's Claim to Privilege. Needham v. Needham and Bennett." *The Times*, June 10, 1921.

"Doctors and Patients." *Daily Chronicle*, June 10, 1921.

"Doctors Must Tell. Secrecy Pledge Overruled by Judge. No Privilege in Law." *Daily Chronicle*, June 10, 1921.

"Doctrine of Privileged Communications." *Journal of the American Medical Association* 54 (1910): 2094.

Donnelly, Laura. "Hospital Records of All NHS Patients Sold to Insurers." *The Telegraph*, February 23, 2014, http://www.telegraph.co.uk/news/health/news/10656893/Hospital-records-of-all-NHS-patients-sold-to-insurers.html.

"Dr. John Elliott." *The Times*, December 20, 1921.

Dunne, A. B. "Medical Certificates Prior to Marriage and Venereal Diseases." *The Lancet* 1918 (October 5): 472.

"Dying Declarations Made after Refusal of Physician to Treat Abortion Case without History—Construction of Amended Statutes with Reference to Privilege and Competency of Witnesses (State vs. Law (Wis.), 136 N. W. R. 803)." *Journal of the American Medical Association* 60 (1913): 1829–30.

Ebermayer, Ludwig. "Die Unruhen in Berlin-Moabit und das Zeugnisverweigerungsrecht der Aerzte." *Ärztliches Vereinsblatt für Deutschland* 37 (1910): 828–29.

E. B. P., "Evidence: Privileged Communications to Physicians: Waiver." *California Law Review* 6 (1918): 300–302.

Edler, William. "The Reporting of Venereal Diseases by Physicians." *Journal of the American Medical Association* 74 (1920): 1764–67.

"Entscheidungen des ärztlichen Ehrengerichtshofs. Die ärztliche Schweigepflicht, Geschlechtskrankheiten." *Berliner Aerzte-Correspondenz* 13 (1908): 55–56.

Entscheidungen des Preußischen Ehrengerichtshofes für Ärzte, vol. 1. Berlin: Richard Schoetz, 1908.

Entscheidungen des Preußischen Ehrengerichtshofes für Ärzte, vol. 4. Berlin: Richard Schoetz, 1927.

Entscheidungen des Reichsgerichts in Civilsachen 53 (1903): 315–19 (decision of January 19, 1903).

Entscheidungen des Reichsgerichts in Strafsachen 13 (1886): 60–64 (decision of October 22, 1885).

Entscheidungen des Reichsgerichts in Strafsachen 19 (1889): 364–67 (decision of July 8, 1889).

Entscheidungen des Reichsgerichts in Strafsachen 38 (1905): 62–66 (decision of May 16, 1905).

Entscheidungen des Reichsgerichts in Strafsachen 48 (1914): 269–74 (decision of May 16, 1914).

"Entwurf eines Gesetzes, betreffend die Bekämpfung gemeingefährlicher Krankheiten." In *Stenographische Berichte des Deutschen Reichstages, 10. Legislaturperiode, 1. Session, 1898–1900*, supplement vol. 6: 4189–93.

"Ethics of Professional Secrets." *Journal of the American Medical Association* 35 (1900): 362–63.

Evans, J. Howell. "The Medico-Legal Significance of Gonorrhoea." *Transactions of the Medico-Legal Society* 5 (1907/8): 155–59.

Exner, Bjarne. "Das Berufsgeheimnis des Arztes gemäß § 300 des Str. G. B." (Jur. Diss., Ruprecht-Karls-Universität Heidelberg, 1909).

"Extent of Waiver of Privilege by Bringing Action for Malpractice." *Journal of the American Medical Association* 50 (1908): 1217–18.

Field, G. W., and John B. Uhle. "Privileged Communications." *The American Law Register* 37 (1889): 1–21.

Flesch, Max. "Das ärztliche Berufsgeheimnis und die Bekämpfung der Geschlechtskrankheiten." *Zeitschrift für Bekämpfung der Geschlechtskrankheiten* 4 (1905): 32–51.

Flint, Austin. *Medical Ethics and Etiquette: The Code of Ethics Adopted by the American Medical Association, with Commentaries*. New York: D. Appleton and Company, 1893.

Forster, Wilhelm. "Das ärztliche Berufsgeheimnis nach dem Recht der Reichsärzteordnung und seine Geltung im Recht der Reichsversicherungsordnung" (Jur. Diss., Leopold-Franzens-Universität zu Innsbruck, 1939).

Friedersdorff, Ernst. "Die unbefugte Offenbarung von Privatgeheimnissen durch Rechtsbeistände, Medizinalpersonen und ihre Gehülfen. § 300 St.G. B." (Jur. Diss., Friedrichs-Universität Halle-Wittenberg, 1906).

Fromme. *Die rechtliche Stellung des Arztes und seine Pflicht zur Verschwiegenheit im Beruf* (= *Berliner Klinik*, No. 165). Berlin: Fischer, 1902.

Gans, Richard Otto. "Das ärztliche Berufsgeheimnis des § 300 RStrGB" (Jur. Diss., Ruprecht-Karls-Universität Heidelberg, 1907).

Gilman, Charlotte Perkins. *The Crux* [1911], ed. and with an intro. by Jennifer S. Tuttle. Newark: University of Delaware Press, 2002.

Glaister, John. *A Text-Book of Medical Jurisprudence and Toxicology* [1st ed. 1902]. 4th ed. Edinburgh: E. & S. Livingstone, 1921.

Godson, Clement. "Retained Placenta." *British Medical Journal* 1896: 873 (April 4).

Graham Little, E. "Medical Privilege: Doctors and the Courts." *The Times*, November 14, 1927.

Gransee, Gerhard. "Das Berufsgeheimnis und sein strafrechtlicher Schutz im geltenden Recht und in den Entwürfen" (Jur. Diss., Universität Leipzig, 1928).

Greene, Frances M. "Reportability of Syphilis and Gonorrhea." *Journal of the American Medical Association* 57 (1911): 1049–52.

Gregory, John. *Lectures on the Duties and Qualifications of a Physician*, new edition, corrected and enlarged. London: W. Strahan and T. Cadell, 1772.

H., Dr. "Ueber Verschwiegenheitspflicht und Zeugnisverweigerungsrecht des Arztes vor Gericht." *Aerztliches Vereinsblatt für Deutschland* 30 (1903): 248–51.

Hale, Edwin M. *The Great Crime of the Nineteenth Century*. Chicago: C. S. Halsey, 1867.

Hamilton, John B. "'Medical' Legislation and How to Obtain It." *Journal of the American Medical Association* 28 (1897): 1005–6.

Hays, Isaac. "Code of Ethics" [1847]. In *The Codification of Medical Morality*, vol. 2: *Anglo-American Medical Ethics and Medical Jurisprudence in the Nineteenth Century*, ed. Robert Baker, 75–87. Dordrecht: Kluwer Academic, 1995.

Heger, Alfred. "Berufsgeheimnis und Abtreibung" (Med. Diss., Julius-Maximilians-Universität Würzburg, 1940).

Heinemann, Hugo. "Das ärztliche Berufsgeheimnis." *Deutsche Medizinische Wochenschrift* 26 (1905): 1037–38.

Hellwig, Albert. "Die Bedeutung der Reichsärzteordnung für die Rechtspflege." *Deutsche Justiz* 98 (1936): 370–73.

———. "Die Beschlagnahme ärztlicher Krankenjournale nach geltendem und künftigem Recht." *Deutsche Medizinische Wochenschrift* 36 (1910): 2152–53.

———. "Die civilrechtliche Bedeutung der Geschlechtskrankheiten." *Zeitschrift für Bekämpfung der Geschlechtskrankheiten* 1 (1903): 26–63.

———. "Die Neuregelung des Berufsgeheimnisses des Arztes." *Deutsche Medizinische Wochenschrift* 62 (1936): 153–54.

Herzog, Alfred W. *Medical Jurisprudence*. Indianapolis: Bobbs-Merrill, 1931.

Hicks, John Braxton. "Notes of Cases in Obstetric Jurisprudence." *The Lancet* 126 (1885): 285–86.

Hilgers, Wilhelm. "Schweigepflicht und Zeugnisverweigerungsrecht des Arztes" (Jur. Diss., Rheinische Friedrich-Wilhelms-Universität Bonn, 1930).

Hodder, Lord. "An Address on Eugenics and the Doctor." *British Medical Journal* 3805 (1933): 1057–60.

Hoerning. "Berufsgeheimniss." *Deutsche Medicinische Wochenschrift* 23 (1903): 415–16.

Holton, Henry D., Leartus Connor, Daniel T. Nelson, and Benjamin Lee. "Code of Medical Ethics and Etiquette of the American Medical Association. Report of Majority Committee." *Journal of the American Medical Association* 22 (1894): 507–10.

Howell, Thomas Bayly. *A Complete Collection of State Trials*, vol. 20. London: Longman, Hurst, Rees etc., 1816.

"Human Sterilization in Germany and the United States" (Editorial). *Journal of the American Medical Association* 102 (1934): 1501–2.

Hurty, John N. "The President's Address." *American Journal of Public Health* 2 (1912): 753–65.

"Infringement of Professional Secrecy." *Journal of the American Medical Association* 44 (1905): 803.

"An Interesting Case of Medical Secrecy." *The Lancet* 1921 (April 16): 822.

Jacobs, Samuel I. "Evidence: Privileged Communications between Physician and Patient in California: Cal. Code Civ. Proc. § 1881 (4)." *California Law Review* 20 (1932): 302–11.

Jaeger. "Das Berufsgeheimnis der Aerzte und Anwälte." *Deutsche Juristen-Zeitung* 11 (1906): 800–805.

Jellinek, Walter. "Der Umfang der Verschwiegenheitspflicht des Arztes und des Anwalts." *Monatsschrift für Kriminalpsychologie und Strafrechtsreform* 3 (1906/7): 656–93.

"A Judge on Professional Secrecy." *The Lancet* 184 (1914): 1430–31.

"A Judge on Professional Secrecy" (Editorial). *The Lancet* 185 (1915): 28–29.

Jummel, Friedrich Ottomar. "Der § 300 Str.G.B., ein Versuch seiner Auslegung" (Jur. Diss., Universität Leipzig, 1903).

Kitchin, D. Harcourt. *Law for the Medical Practitioner*. London: Eyre and Spottiswoode Ltd., 1941.

"Kitson v. Playfair." *British Medical Journal* 1896: 1161–62 (May 9).

"Kitson v. Playfair." *The Lancet* 147 (1896): 1292–93.

"Kitson v. Playfair and Wife." *British Medical Journal* 1896: 815–19 (March 28), 882–84 (April 4).

"Kitson v. Playfair and Wife." *The Times*, March 23, 25, 26, and 28, 1896.

Kohler, Josef. "Stellung der Rechtsordnung zur Gefahr der Geschlechtskrankheiten." *Zeitschrift für Bekämpfung der Geschlechtskrankheiten* 2 (1903/4): 19–30.

Krahmer, Dr. "Der Zeugnisszwang der Aerzte." *Berliner Klinische Wochenschrift* 1875, nos. 26–27: 365–66, 378.

"Law Report, Jan. 13." *The Times*, January 14, 1920.

Lehmann, Rudolf. "Die Schweigepflicht des Arztes und das Gesetz zur Bekämpfung der Geschlechtskrankheiten." *Die Medizinische Welt* 20 (1928): 776–78.

"Liability of a Physician for Revealing out of Court His Patient's Confidences." *Harvard Law Review* 34 (1920–21): 312–14.

Liebmann, J. *Die Pflicht des Arztes zur Bewahrung anvertrauter Geheimnisse*. Frankfurt a. M.: Joseph Baer & Co., 1886.

"London Letter: Professional Secrecy." *Journal of the American Medical Association* 65 (1915): 634–35.

Lübke, Carl. "Die Schweigepflicht der Rechtsanwälte, Aerzte und ihrer Gehilfen" (Jur. Diss., Universität Jena, 1920).

Mackenzie, Muir. "Memorandum on the Law of the Obligation of Medical Practitioners with Regard to Professional Secrecy." *The Lancet* 153 (1899): 787–88.

Marcus, Dr. "Wegen Verweigerung des Zeugnisses." *Aerztliches Vereinsblatt* 11 (1884): 94–96.

Markham, H. C. "Foeticide and Its Prevention." *Journal of the American Medical Association* 9 (1888): 805–6.

Marshall, C. F. "Venereal Disease and Professional Secrecy." *The Lancet* 195 (1920): 171.

M. D. "Professional Secrecy and the Law." *British Medical Journal* 1896: 871–72 (April 4).

"Medical Confidences." *Journal of the American Medical Association* 33 (1899): 1431.

"Medical Confidences and Medical Testimony." *Journal of the American Medical Association* 31 (1898): 309–10.

"Medical Confidences and Professional Honor." *Journal of the American Medical Association* 26 (1896): 783–84.

"Medical Evidence in Divorce. Effect on Treatment Schemes. Value of Secrecy." *The Times*, January 15, 1920.

"Medical Privilege." *The Times*, April 7, 1896.

"Medical Professional Privilege." *British Medical Journal* 1920: 135–36 (January 24).

"Medical Trial. Queen's Bench Division, High Court of Justice. (Before Mr. Justice Hawkins.) Kitson v. Playfair and Wife." *The Lancet* 147 (1896): 896–97, 961–62.

"Michigan Medical Laws." *Journal of the American Medical Association* 33 (1899): 233.

Mittermaier, Wolfgang. "Das ärztliche Berufsgeheimnis nach der Reichsärzteordnung." *Monatsschrift für Kriminalpsychologie und Strafrechtsreform* 27 (1936): 153–54.

———. "Gutachten über § 300 R.St.G.B." *Zeitschrift für die gesamte Strafrechtswissenschaft* 21 (1901): 197–258.

Moll, Albert. *Ärztliche Ethik: Die Pflichten des Arztes in allen Beziehungen seiner Thätigkeit*. Stuttgart: Ferdinand Enke, 1902.

———. *Ein Leben als Arzt der Seele: Erinnerungen*. Dresden: Carl Reissner Verlag, 1936.

———. "Neuere Fragen zum ärztlichen Berufsgeheimnis." *Berliner Aerzte-Correspondenz* 16 (1911): 1–4.

Morrow, Prince A. *Social Diseases and Marriage: Social Prophylaxis*. New York: Lea Brothers & Co., 1904.

Moser, Oswald. "Das ärztliche Berufsgeheimnis (RAeO. vom 15. XII. 1935)" (Med. Diss., Ernst-Moritz-Arndt-Universität Greifswald, 1936).

Müller-Welt, Christian. "Das Berufsgeheimnis des Arztes und Apothekers nach der Reichsärzte-und Apothekerordnung" (Jur. Diss., Friedrich-Alexanders-Universität zu Erlangen, 1938).

Neisser, Albert. "Abänderung des § 300 des Reichs-Strafgesetzbuches und ärztliches Anzeigerecht in ihrer Bedeutung für die Bekämpfung der Geschlechtskrankheiten." *Zeitschrift für Bekämpfung der Geschlechtskrankheiten* 4 (1905): 1–28.

Neukamp, Franz. "Das ärztliche Berufsgeheimnis nach der Reichsärzteordnung." *Monatsschrift für Kriminalpsychologie und Strafrechtsreform* 28 (1937): 77–84.

———. "Einige Vorschläge für Erb -und Ehegesundheitssachen." *Monatsschrift für Kriminalpsychologie und Strafrechtsreform* 27 (1936): 243–52.

Nuffield Council on Bioethics. *The Collection, Linking and Use of Data in Biomedical Research and Health Care: Ethical Issues* (London 2015), http://nuffieldbioethics.org/wp-content/uploads/Biological_and_health_data_web.pdf.

"The Obligation of Professional Secrecy." *British Medical Journal* 1896: 861–62 (April 4).

"The Obligation of Professional Secrecy. Medical and Legal Opinions. I.—The Legal Issues. [By a Legal Correspondent.]." *British Medical Journal* 1896: 869–70 (April 4).

Olshausen, Justus. *Kommentar zum Strafgesetzbuch für das Deutsche Reich*, 4th revised edition, vol. 2. Berlin: Verlag von Franz Vahlen, 1892.

Ottmer, F. *Schweigen. Erzählung*. Berlin: Concordia Deutsche Verlags-Anstalt, 1902.

Pallaske. "Die Schweigepflicht des Arztes." *Deutsche Juristen-Zeitung* 11 (1906): 294–97.

Percival, Thomas. "Medical Ethics; or a Code of Institutes and Precepts, Adapted to the Professional Conduct of Physicians and Surgeons" (1803). In *Percival's Medical Ethics*, ed. Chauncey D. Leake, 61–205. Huntington, NY: Robert E. Krieger, 1975.

"Physicians and Privileged Communications." *Journal of the American Medical Association* 51 (1908): 1170.

"The Physician's Responsibility: An Important Decision on Professional Secrecy" (Editorial). *Journal of the American Medical Association* 75 (1920): 1207–8.

"Physical Examination and Privilege Waiving." *Journal of the American Medical Association* 48 (1907): 903.

Placzek, Siegfried. "Aerztliches Berufsgeheimnis und Ehe." In *Krankheiten und Ehe. Darstellung der Beziehungen zwischen Gesundheits-Störungen und Ehegemeinschaft*, ed. H. Senator and S. Kaminer, 793–806. Munich: J. F. Lehmann, 1904.

———. *Das Berufsgeheimnis des Arztes*. Leipzig: Verlag von Georg Thieme, 1893.

———. *Das Berufsgeheimnis des Arztes*, 3rd enlarged and revised edition. Leipzig: Verlag von Georg Thieme, 1909.

———. "Medico-Professional Secrecy in Relation to Marriage." In *Marriage and Disease. Being an Abridged Edition of "Health and Disease in Relation to Marriage and the Married State,"* ed. H. Senator and S. Kaminer, 438–40. London: Rebman Limited, 1907.

"The Privacy of Venereal Clinics." *The Lancet* 195 (1920): 163.

"Privilege—a Medical Difficulty." *The Times*, April 4, 1896.

"Privilege Attaches to Hospital Records—When Treated as Public Records." *Journal of the American Medical Association* 79 (1922): 2188.

"Privileged Communications." *Journal of the American Medical Association* 62 (1914): 1350–51.

"Privileged Communications to Physicians." *Yale Law Journal* 30 (1921): 289–91.

"Privileged Communications under Amended Statute." *Journal of the American Medical Association* 64 (1915): 1446.

"Privilege Is That of Patient and Not of Physician." *Journal of the American Medical Association* 73 (1919): 1083.

"Privilege Not Waived by Bringing Action for Personal Injuries." *Journal of the American Medical Association* 50 (1908): 723–24.

"Privilege Not Waived nor Statute Modified." *Journal of the American Medical Association* 60 (1913): 1830.

"Privilege Waived by Bringing of Action for Malpractice—Evidence of Defense by Medical Society Not Admissible." *Journal of the American Medical Association* 58 (1912): 511.

"Professional Communications to Physicians Should Be Privileged." *Journal of the American Medical Association* 28 (1897): 374.

"Professional Confidences." *The Lancet* 147 (1896): 1524.

"Professional Secrecy." *Journal of the American Medical Association* 35 (1900): 104.

"Professional Secrecy." *Journal of the American Medical Association* 44 (1905): 1865.

"Professional Secrecy." *The Lancet* 148 (1896): 254.

"Professional Secrecy." *The Lancet* 181 (1913): 52.

"Professional Secrecy of Doctors. Mr. Justice McCardie on Duty to Court." *The Times*, July 19, 1927.

Purrington, William A. "An Abused Privilege." *Columbia Law Review* 6 (1906): 388–422.

———. "Of Certain Legal Relations of Physicians and Surgeons to Their Patients and to One Another." In *A System of Legal Medicine*, 2nd ed., ed. Allan McLane Hamilton and Lawrence Godkin, vol. 1, 595–648a. New York: E. B. Treat and Co., 1900.

———. "Professional Secrecy and the Obligatory Notification of Venereal Diseases." *New York Medical Journal* 85 (1907): 1206–10.

———. *A Review of Recent Legal Decisions Affecting Physicians, Dentists, Druggists and the Public Health*. New York: E. B. Treat and Co., 1899.

"Regulation of Venereal Disease." *Journal of the American Medical Association* 39 (1902): 776.

Riddell, [George Allardice] Lord. *Medico-Legal Problems*. London: H. K. Lewis and Co., 1929.

Robertson, W. G. Aitchison. *Medical Conduct and Practice: A Guide to the Ethics of Medicine*. London: A. and C. Black, 1921.

Rudeck, Wilhelm. *Medizin und Recht: Geschlechtsleben und–Krankheiten in medizinisch-juristisch-kulturgeschichtlicher Bedeutung*, 2nd ed. Berlin: H. Barsdorf, 1902.

"Rule as to Privileged Communications Applied to Roentgenologist." *Journal of the American Medical Association* 71 (1918): 2019.

"Runderlaß des Reichs -und Preußischen Ministers des Innern, betr. Schweigepflicht bei Durchführung des Gesetzes zur Verhütung erbkranken Nachwuchses. Vom 31. Mai 1935." *Ärzteblatt für Berlin* 40 (1935): 257.

Sanderson, William, and E. B. A. Rayner. *An Introduction to the Law and Tradition of Medical Practice*. London: H. K. Lewis & Co., 1926.

Sandheim, Hugo. "Die unbefugte Offenbarung von Privatgeheimnissen nach § 300 St. G. B." (Jur. Diss., Friedrichs-Universität Halle-Wittenberg, 1904).

Saundby, Robert. *Medical Ethics: A Guide to Professional Conduct*, 2nd ed. London: Charles Griffin & Company, 1907.

Sauter, Fritz. *Das Berufsgeheimnis und sein strafrechtlicher Schutz. (§ 300 R.St.G.B.)*. Breslau: Schletter'sche Buchhandlung, 1910.

Schlegtendal, Dr. "Das Berufsgeheimniss der Aerzte." *Deutsche Medicinische Wochenschrift* 21 (1895): 503–6.

Schmalzl, Hans. "Der Schutz des Privatgeheimnisses. Ein Vergleich zwischen § 300 St.G.B. und § 293 Entwurf 1925 mit Anhang: Der Schutz des Privatgeheimnisses nach § 325 Entwurf 1927" (Jur. Diss., Friedrich-Alexanders-Universität zu Erlangen, 1928).

Schmidt, Heinrich. *Das ärztliche Berufsgeheimnis* (Jur. Diss., Universität Leipzig). Jena: Gustav Fischer, 1907.

Schmitz, Wilhelm. "Reichsärzteordnung und Berufsgeheimnis." *Die Medizinische Welt* 10 (1936): 757–58.

———. "Sterilisierungsgesetz und ärztliches Berufsgeheimnis." *Die Medizinische Welt* 9 (1935): 205–6.

Schumacher, Willy. *Das ärztliche Berufsgeheimnis nach § 300 RStGB.* Berlin: Verlagsbuchhandlung von Richard Schoetz, 1931.

Schwerdtfeger, Richard. "Die Verletzung des Berufsgeheimnisses nach § 300 RStGB" (Jur. Diss., Friedrich-Alexanders-Universität zu Erlangen, 1926).

S. C. S. "Family Duties and Professional Secrecy." *British Medical Journal* 1896: 872–73 (April 4).

"The Seal of Professional Secrecy." *The Lancet* 155 (1900): 1292–93.

Seaman, Allen H., Charles R. Brock, Edmund J. A. Rogers, H. G. Wetherill, and H. H. Hawkins. "Symposium. Criminal Abortion. The Colorado Law on Abortion." *Journal of the American Medical Association* 40 (1903): 1096–99.

"Second German Preventive Congress." *Journal of the American Medical Association* 44 (1905): 1207.

"Secrecy and Privilege." *Medical Press and Circular* 1896: 402–3 (April 15).

Seréxhe, Léon. "Die Verletzung fremder Geheimnisse" (Jur. Diss., Universität Freiburg i. B., 1906).

Simonson. "Das Berufsgeheimnis der Aerzte und deren Recht der Zeugnisverweigerung." *Deutsche Juristen-Zeitung* 9 (1904): 1014–17.

"Some Pathological Features." *British Medical Journal* 1896: 870 (April 4).

Spengemann, Walter. "Das Berufsgeheimnis des Arztes" (Med. Diss., Westfälische Wilhelms-Universität Münster, 1928).

[Sprigge, Samuel Squire]. *The Conduct of Medical Practice*, ed. the Editor of "The Lancet" and Expert Collaborators. London: The Lancet, 1927.

"Statutes Relative to Privileged Communications and Vital Statistics." *Journal of the American Medical Association* 79 (1922): 325.

Stebbins, Emma. *Charlotte Cushman: Her Letters and Memories of Her Life*. Boston: Houghton, Osgood and Company, 1879.

Stenographische Berichte über die Verhandlungen des Preußischen Hauses der Abgeordneten, 20. Legislaturperiode, I. Session 1904/05, vol. 6 and 8. Berlin: W. Moeser, 1905.

Stenographische Berichte über die Verhandlungen des Preußischen Herrenhauses in der Session 1905/06. Berlin: Julius Sittenfeld, 1906.

Strock, Daniel. "A Plea for the Physician on the Witness Stand." *Transactions of the Medical Society of New Jersey* 1901: 169–77.

Stryker, Lloyd Paul. *Courts and Doctors*. New York: Macmillan, 1932.

Taeusch, C. F. "Should the Doctor Testify?" *International Journal of Ethics* 38 (1928): 401–15.

Tait, William C. "The Physician's Obligation to Secrecy." *American Medicine* 4 (1902): 265–67.

———. "Professional Secrecy and Its Legal Aspects." *Journal of the American Medical Association* 33 (1899): 458–62.

Thibierge, Georges. *Syphilis et Déontologie*. Paris: Masson et Cie, 1903.

The Trial of Elizabeth Duchess Dowager of Kingston for Bigamy, Before the Right Honourable The House of Peers, in Westminster-Hall, in Full Parliament . . . Published by Order of the House of Peers. London: Charles Bathurst, 1776.

Trost, Hermann. "Das ärztliche Berufsgeheimnis nach § 13 der Reichsärzteordnung" (Jur. Diss., Universität Rostock, 1940).

Vogler, Eduard. "Betrachtungen zur Begrenzung der ärztlichen Schweigepflicht nach § 13 Abs. 3 der Reichsärzteordnung" (Jur. Diss., Philipps-Universität zu Marburg, 1939).

Vollmann. "Umschau. Beratungstellen—Berufsgeheimnis—Kurpfuscherei—ärztliche Ausbildung." *Ärztliches Vereinsblatt für Deutschland* 43 (1916): 455–62.

Wagner, Gerhard. "Die Reichsärzteordnung ein Instrument nationalsozialistischer Gesundheitspolitik." *Deutsches Ärzteblatt* 65 (1935): 1233–34.

"Waiver of Privilege by Plaintiffs." *Journal of the American Medical Association* 79 (1922): 2250.

"Waiver of Privilege in Personal Injury Case." *Journal of the American Medical Association* 80 (1923): 1172.

"Waivers of Privilege." *Journal of the American Medical Association* 57 (1911): 413.

Warren, Samuel D., and Louis D. Brandeis. "The Right to Privacy." *Harvard Law Review* 4 (1890): 193–220.

Weizmann, Hans. "Das Berufsgeheimnis. (§ 300 RStGB.)" (Jur. Diss., Universität Breslau, 1909).

"When Doctrine of 'Res Ipsa Loquitur' Applies." *Journal of the American Medical Association* 91 (1928): 1919–20.

Wigmore, John Henry. *A Treatise on the System of Evidence in Trials at Common Law*, vol. 4. Boston: Little, Brown, and Company, 1905.

Willcox, W. H. "Criminal Abortion and Professional Secrecy." *The Lancet* 185 (1915): 97–98.

Williams, Campbell. "A Lecture on the Ethics of the Medical Profession in Relation to Syphilis and Gonorrhoea. Delivered before the Harveian Society on Feb. 8th, 1906." *The Lancet* 167 (1906): 361–63.

"Wisconsin Doctrine as to Privileged Communications." *Journal of the American Medical Association* 47 (1904): 1577.

Woods, Hugh. "Medical Secrecy." *The Lancet* 203 (1924): 853–54.

Woodward, William C. "A Brief Statement of the Principles Underlying the Physician's Obligation to Secrecy." *American Medicine* 4 (1902): 738–40.

Zschok, Walther. "§ 300 StrGB" (Jur. Diss., Universität Rostock, 1903).

Secondary Literature

Baker, Robert. *Before Bioethics: A History of American Medical Ethics from the Colonial Period to the Bioethics Revolution*. New York: Oxford University Press, 2013.

———. "Deciphering Percival's Code." In *The Codification of Medical Morality*, vol. 1: *Medical Ethics and Etiquette in the Eighteenth Century*, ed. Robert Baker, Dorothy Porter, and Roy Porter, 179–211. Dordrecht: Kluwer Academic, 1993.

———. "The Historical Context of the American Medical Association's 1847 *Code of Ethics*." In *The Codification of Medical Morality*, vol. 2: *Anglo-American Medical Ethics and Medical Jurisprudence in the Nineteenth Century*, ed. Robert Baker, 47–63. Dordrecht: Kluwer Academic, 1995.

Baldwin, Peter. *Contagion and the State in Europe, 1830–1930*. Cambridge: Cambridge University Press, 2005.

Benzenhöfer, Udo. *Zur Genese des Gesetzes zur Verhütung erbkranken Nachwuchses*. Münster: Klemm & Oelschläger, 2006.

Bernstein, Frances L. "Behind the Closed Door: VD and Medical Secrecy in Early Soviet Medicine." In *Soviet Medicine: Culture, Practice, and Science*, ed. F. L. Bernstein, Christopher Burton and Dan Healy, 92–110. DeKalb: Northern Illinois University Press, 2010.

Bland, Lucy, and Lesley A. Hall. "Eugenics in Britain: The View from the Metropole." In *The Oxford Handbook of the History of Eugenics*, ed. Alison Bashford and Philippa Levine, 213–27. Oxford: Oxford University Press, 2010.

Brandt, Alan. *No Magic Bullet: A Social History of Venereal Disease in the United States since 1880*. New York: Oxford University Press, 1985.

British Medical Association. *Medical Ethics Today: The BMA's Handbook of Ethics and Law*, 2nd ed. London: BMJ Books, 2004.

Brookes, Barbara. *Abortion in England 1900–1967*. London: Croom Helm, 1988.

Connelly, Mark Thomas. "Prostitution, Venereal Disease, and American Medicine." In *Women and Health in America: Historical Readings*, ed. Judith Walzer Leavitt, 196–221. Madison: University of Wisconsin Press, 1984.

Cooke, A. M. *A History of the Royal College of Physicians of London*, vol. 3. Oxford: Clarendon Press, 1972.

Czarnowski, Gabriele. "Women's Crimes, State Crimes: Abortion in Nazi Germany." In *Gender and Crime in Modern Europe*, ed. Margaret L. Arnot and Cornelie Usborne, 238–56. London: UCL Press, 1999.

Davidson, Roger. *Dangerous Liaisons: A Social History of Venereal Disease in Twentieth-Century Scotland*. Amsterdam: Rodopi, 2000.

Davidson, Roger, and Lutz D. H. Sauerteig. "Law, Medicine and Morality: A Comparative View of Twentieth-Century Sexually Transmitted Disease Controls." In *Coping with Sickness: Medicine, Law and Human Rights—Historical Perspectives*, ed. John Woodward and Robert Jütte, 127–47. Sheffield: European Association for the History of Medicine and Health Publications, 2000.

DeWitt, Clinton. *Privileged Communications between Physician and Patient*. Springfield: Charles C. Thomas, 1958.

Eichelbrönner, Nicolas. *Die Grenzen der Schweigepflicht des Arztes und seiner berufsmäßig tätigen Gehilfen nach § 203 StGB im Hinblick auf Verhütung und Aufklärung von Straftaten: Anzeigepflichten, Auskunftspflichten und Offenbarungsbefugnisse gegenüber den Strafverfolgungsbehörden* (Jur. Diss., Universität Würzburg, 2001). Frankfurt/Main: Peter Lang, 2001.

Fairchild, Amy L., Ronald Bayer, and James Colgrove, with Daniel Wolfe. *Searching Eyes: Privacy, the State, and Disease Surveillance in America*. Berkeley: University of California Press, 2007.

Ferguson, Angus H. "Exploring the Myth of a Scottish Privilege: A Comparison of the Early Development of the Law on Medical Confidentiality in Scotland and England." In *Medicine, Law and Public Policy in Scotland c. 1850–1990. Essays Presented to Anne Crowther*, ed. Mark Freeman, Eleanor Gordon, and Krista Maglen, 125–40. Dundee: Dundee University Press, 2011.

———. "The Lasting Legacy of a Bigamous Duchess: The Benchmark Precedent for Medical Confidentiality." *Social History of Medicine* 19 (2006): 37–53.

———. "Medical Confidentiality in the Military." In *Military Medical Ethics for the 21st Century*, ed. Michael L. Gross, and Don Carrick, 209–24. Farnham: Ashgate, 2013.

———. "The Role of History in Debates Regarding the Boundaries of Medical Confidentiality and Privacy." *Journal of Medical Law and Ethics* 3 (2015): 65–81.

———. *Should a Doctor Tell? The Evolution of Medical Confidentiality in Britain*. Farnham: Ashgate, 2013.

———. "Speaking Out about Staying Silent: An Historical Examination of Medico-Legal Debates over the Boundaries of Medical Confidentiality." In *Lawyers' Medicine: The Legislature, the Courts and Medical Practice, 1760–2000*, ed. Imogen Goold and Catherine Kelly, 99–124. Oxford: Hart, 2009.

Gervat, Claire. *Elizabeth: The Scandalous Life of the Duchess of Kingston*. London: Century, 2003.

Grossmann, Atina. *Reforming Sex: The German Movement for Birth Control and Abortion Reform, 1920–1950*. New York: Oxford University Press, 1995.

Hall, Lesley A. "Venereal Diseases and Society in Britain, from the Contagious Diseases Acts to the National Health Service." In *Sex, Sin and Suffering: Venereal Disease in European Society since 1870*, ed. Roger Davidson and Lesley A. Hall, 120–36. London: Routledge, 2001.

Huerkamp, Claudia. *Der Aufstieg der Ärzte im 19. Jahrhundert. Vom gelehrten Stand zum professionellen Experten: Das Beispiel Preußens*. Göttingen: Vandenhoeck and Ruprecht, 1985.

Jay, Rosemary. *Data Protection Law and Practice*, 4th ed., 1st suppl. Andover: Sweet and Maxwell, Thomson Reuters, 2014.

Kelly, Catherine, and Imogen Goold. "Introduction: Lawyers' Medicine: The Interaction of the Medical Profession and the Law, 1760–2000." In *Lawyers'*

Medicine: The Legislature, the Courts and Medical Practice, 1760–2000, ed. Imogen Goold and Catherine Kelly, 1–15. Oxford: Hart, 2009.

Keown, John. *Abortion, Doctors and the Law: Some Aspects of the Legal Regulation of Abortion in England from 1803 to 1982*. Cambridge: Cambridge University Press, 1988.

Kernahan, Peter J. "'A Condition of Development': Muckrakers, Surgeons, and Hospitals, 1890–1920." *Journal of the American College of Surgeons* 206 (2008): 376–84.

Kevles, Daniel J. *In the Name of Eugenics: Genetics and the Uses of Human Heredity*. Berkeley: University of California Press, 1985.

Kline, Wendy. "Eugenics in the United States." In *The Oxford Handbook of the History of Eugenics*, ed. Alison Bashford and Philippa Levine, 511–22. Oxford: Oxford University Press, 2010.

Lamprecht, Rolf. "Wieviel ist das Arztgeheimnis noch wert? Zur Güterabwägung zwischen Privatsphäre und Strafrechtspflege, erläutert am Memminger Exempel." *Zeitschrift für Rechtspolitik* 22, no. 8 (1989): 290–93.

Lang, Franziska. *Das Recht auf informationelle Selbstbestimmung des Patienten und die ärztliche Schweigepflicht in der gesetzlichen Krankenversicherung*. Baden-Baden: Nomos Verlagsgesellschaft, 1997.

Leavitt, Judith Walzer. *Typhoid Mary: Captive to the Public's Health*. Boston: Beacon Press, 1996.

Macnicol, John. "Eugenics and the Campaign for Voluntary Sterilization in Britain between the Wars." *Social History of Medicine* 2 (1989): 147–69.

Maehle, Andreas-Holger. *Doctors, Honour and the Law: Medical Ethics in Imperial Germany*. Basingstoke: Palgrave Macmillan, 2009.

———. "'God's Ethicist': Albert Moll and His Medical Ethics in Theory and Practice." *Medical History* 56 (2012): 217–36.

———. "Preserving Confidentiality or Obstructing Justice? Historical Perspectives on a Medical Privilege in Court." *Journal of Medical Law and Ethics* 3 (2015): 91–108.

———. "Protecting Patient Privacy or Serving Public Interests? Challenges to Medical Confidentiality in Imperial Germany." *Social History of Medicine* 16 (2003): 383–401.

Maehle, Andreas-Holger, and Sebastian Pranghofer. "Medical Confidentiality in the Late Nineteenth and Early Twentieth Centuries: An Anglo-German Comparison." *Medizinhistorisches Journal* 45 (2010): 189–221.

McCullough, Laurence. *John Gregory and the Invention of Professional Medical Ethics and the Profession of Medicine*. Dordrecht: Kluwer Academic, 1998.

McHale, Jean. "From *X v Y* to care.data and Beyond: Health Care Confidentiality and Privacy in the C21st: A Critical Turning Point?" *Journal of Medical Law and Ethics* 3 (2015): 109–33.

McLaren, Angus. "Privileged Communications: Medical Confidentiality in Late Victorian Britain." *Medical History* 37 (1993): 129–47.

Mendelson, Danuta. "The Duchess of Kingston's Case, the Ruling of Lord Mansfield and Duty of Medical Confidentiality in Court." *International Journal of Law and Psychiatry* 35 (2012): 480–89.

Michalowski, Sabine. *Medical Confidentiality and Crime*. Aldershot: Ashgate, 2003.

Miles, Steven H. *The Hippocratic Oath and the Ethics of Medicine*. Oxford: Oxford University Press, 2004.

Mohr, James C. *Abortion in America: The Origins and Evolution of National Policy, 1800–1900*. New York: Oxford University Press, 1978.

———. "American Medical Malpractice Litigation in Historical Perspective." *Journal of the American Medical Association* 283 (2000): 1731–37.

———. *Licensed to Practice: The Supreme Court Defines the American Medical Profession*. Baltimore: Johns Hopkins University Press, 2013.

———. "Patterns of Abortion and the Response of American Physicians, 1790–1930." In *Women and Health in America: Historical Readings*, ed. Judith Walzer Leavitt, 117–23. Madison: University of Wisconsin Press, 1984.

Mooij, Annet. *Out of Otherness: Characters and Narrators in the Dutch Venereal Disease Debates 1850–1990*. Amsterdam: Rodopi, 1998.

Mooney, Graham. "Public Health versus Private Practice: The Contested Development of Compulsory Infectious Disease Notification in Late-Nineteenth-Century Britain." *Bulletin of the History of Medicine* 73 (1999): 238–67.

Morrice, Andrew A. G. "'Should the Doctor Tell?' Medical Secrecy in Early Twentieth-Century Britain." In *Medicine, Health and the Public Sphere in Britain, 1600–2000*, ed. Steve Sturdy, 60–82. London: Routledge, 2002.

Pattinson, Shaun, and Deryck Beyleveld. "Confidentiality and Data Protection." In *Principles of Medical Law*, 3rd ed., ed. Andrew Grubb, Judith Laing, Jean McHale, 651–723. Oxford: Oxford University Press, 2010.

Polianski, Igor J. *Das Schweigen der Ärzte: Eine Kulturgeschichte der sowjetischen Medizin und ihrer Ethik*. Stuttgart: Franz Steiner Verlag, 2015.

Pratt, Walter F. *Privacy in Britain*. Lewisburg: Bucknell University Press, 1979.

Reagan, Leslie J. *When Abortion Was a Crime: Women, Medicine, and Law in the United States, 1867–1973*. Berkeley: University of California Press, 1997.

Rieger, H.-J. "Zur geschichtlichen Entwicklung der ärztlichen Schweigepflicht." *Deutsche Medizinische Wochenschrift* 100 (1975): 1867–68.

Sauerteig, Lutz D. H. "'The Fatherland Is in Danger, Save the Fatherland!' Venereal Disease, Sexuality and Gender in Imperial and Weimar Germany." In *Sex, Sin and Suffering: Venereal Disease in European Society since 1870*, ed. Roger Davidson and Lesley A. Hall, 76–92. London: Routledge, 2001.

———. *Krankheit, Sexualität, Gesellschaft: Geschlechtskrankheiten und Gesund-*

heitspolitik in Deutschland im 19. und frühen 20. Jahrhundert. Stuttgart: Franz Steiner Verlag, 1999.

Schmuhl, Hans-Walter. *Rassenhygiene, Nationalsozialismus, Euthanasie: Von der Verhütung zur Vernichtung "lebensunwerten Lebens", 1890–1945*, 2nd ed. Göttingen: Vandenhoeck & Ruprecht, 1992.

Seidler, Eduard. "Das 19. Jahrhundert. Zur Vorgeschichte des Paragraphen 218." In *Geschichte der Abtreibung. Von der Antike bis zur Gegenwart*, ed. Robert Jütte, 120–39. Munich: Verlag C. H. Beck, 1993.

Shuman, Daniel W. "The Origins of the Physician-Patient Privilege and Professional Secret." *Southwestern Law Journal* 39 (1985/86): 661–87.

Slovenko, Ralph. *Psychotherapy and Confidentiality. Testimonial Privileged Communication, Breach of Confidentiality, and Reporting Duties*. Springfield: Charles C. Thomas, 1998.

Smith, Russell G. *Medical Discipline. The Professional Conduct Jurisdiction of the General Medical Council, 1858–1990*. Oxford: Clarendon Press, 1994.

Soljak, Michael. "Big Voice or Big Data? The Difficult Birth of care.data." *Journal of Medical Law and Ethics* 3 (2015): 135–42.

Soloway, Richard A. *Demography and Degeneration: Eugenics and the Declining Birthrate in Twentieth-Century Britain*. Chapel Hill: University of North Carolina Press, 1995.

Staufer, Andreas Michael. *Ludwig Ebermayer: Leben und Werk des höchsten Anklägers in der Weimarer Republik unter besonderer Berücksichtigung seiner Tätigkeit im Medizin-und Strafrecht*. Leipzig: Leipziger Universitätsverlag, 2010.

Steinberg-Copek, Jutta. "Berufsgeheimnis und Aufzeichnungen des Arztes im Strafverfahren" (Jur. Diss., Freie Universität Berlin, 1968).

Usborne, Cornelie. *Cultures of Abortion in Weimar Germany*. New York: Berghahn Books, 2007.

———. *The Politics of the Body in Weimar Germany: Women's Reproductive Rights and Duties*. Ann Arbor: University of Michigan Press, 1992.

Van Ingen, Philip. *The New York Academy of Medicine: Its First Hundred Years*. New York: Columbia University Press, 1949.

Villey, Raymond. *Histoire du Secret Médical*. Paris: Seghers, 1986.

Wacks, Raymond. *Privacy and Media Freedom*. Oxford: Oxford University Press, 2013.

Wahrig, Bettina, and Werner Sohn (eds.). *Zwischen Aufklärung, Policey und Verwaltung: Zur Genese des Medizinalwesens 1750–1850*. Wiesbaden: Harrassowitz Verlag, 2003.

Warner, John Harley. "The 1880s Rebellion against the AMA Code of Ethics: 'Scientific Democracy' and the Dissolution of Orthodoxy." In *The American Medical Ethics Revolution. How the AMA's Code of Ethics Has Transformed Physicians' Relationships to Patients, Professionals, and Society*, ed. Robert B. Baker,

Arthur L. Caplan, Linda L. Emanuel, and Stephen R. Latham, 52–69. Baltimore: Johns Hopkins University Press, 1999.

Weindling, Paul. *Health, Race and German Politics between National Unification and Nazism, 1870–1945*. Cambridge: Cambridge University Press, 1989.

Whooley, Owen. *Knowledge in the Time of Cholera: The Struggle over American Medicine in the Nineteenth Century*. Chicago: University of Chicago Press, 2013.

Wichmann, Hermann. *Das Berufsgeheimnis als Grenze des Zeugenbeweises. Ein Beitrag zur Lehre von den Beweisverboten* (Jur. Diss., Universität Göttingen, 2000). Frankfurt/Main: Peter Lang, 2000.

Index